History of Mathematics

Clio Mathematicæ
The Muse of Mathematical Historiography

Craig Smoryński

History of Mathematics

A Supplement

 Springer

Craig Smoryński
429 S. Warwick
Westmont, IL 60559
USA
smorynski@sbcglobal.net

ISBN 978-1-4419-2593-0   e-ISBN 978-0-387-75481-9

Mathematics Subject Classification (2000): 01A05 51-Axx, 15-xx

© 2008 Springer Science+Business Media, LLC
Softcover reprint of the hardcover 1st edition 2008

springer.com

Contents

1 Introduction 1
   1   An Initial Assignment 1
   2   About This Book 7

2 Annotated Bibliography 11
   1   General Remarks 11
   2   General Reference Works 18
   3   General Biography 19
   4   General History of Mathematics 21
   5   History of Elementary Mathematics 23
   6   Source Books 25
   7   Multiculturalism 27
   8   Arithmetic 28
   9   Geometry 28
  10   Calculus 29
  11   Women in Science 30
  12   Miscellaneous Topics 35
  13   Special Mention 36
  14   Philately 38

3 Foundations of Geometry 41
   1   The Theorem of Pythagoras 41
   2   The Discovery of Irrational Numbers 49
   3   The Eudoxian Response 59
   4   The Continuum from Zeno to Bradwardine 67
   5   Tiling the Plane 76
   6   Bradwardine Revisited 83

4 The Construction Problems of Antiquity 87
   1   Some Background 87
   2   Unsolvability by Ruler and Compass 89

3   Conic Sections 100
4   Quintisection 110
5   Algebraic Numbers 118
6   Petersen Revisited 122
7   Concluding Remarks 130

5   **A Chinese Problem** 133

6   **The Cubic Equation** 147
1   The Solution 147
2   Examples 149
3   The Theorem on the Discriminant 151
4   The Theorem on the Discriminant Revisited 156
5   Computational Considerations 160
6   One Last Proof 171

7   **Horner's Method** 175
1   Horner's Method 175
2   Descartes' Rule of Signs 196
3   De Gua's Theorem 214
4   Concluding Remarks 222

8   **Some Lighter Material** 225
1   North Korea's Newton Stamps 225
2   A Poetic History of Science 229
3   Drinking Songs 235
4   Concluding Remarks 241

A   **Small Projects** 247
1   Dihedral Angles 247
2   Inscribing Circles in Right Triangles 248
3    $\cos 9°$ 248
4   Old Values of π 249
5   Using Polynomials to Approximate π 254
6   π *à la* Horner 256
7   Parabolas 257
8   Finite Geometries and Bradwardine's Conclusion 38 257
9   Root Extraction 260
10   Statistical Analysis 260
11   The Growth of Science 261
12   Programming 261

Index 263

1

Introduction

1 An Initial Assignment

I haven't taught the history of mathematics that often, but I do rather like the course. The chief drawbacks to teaching it are that i. it is a lot more work than teaching a regular mathematics course, and ii. in American colleges at least, the students taking the course are not mathematics majors but education majors— and and in the past I had found education majors to be somewhat weak and unmotivated. The last time I taught the course, however, the majority of the students were graduate education students working toward their master's degrees. I decided to challenge them right from the start:

Assignment.[1] In *An Outline of Set Theory*, James Henle wrote about mathematics:

> Every now and then it must pause to organize and reflect on what it is and where it comes from. This happened in the sixth century B.C. when Euclid thought he had derived most of the mathematical results known at the time from five postulates.

Do a little research to find as many errors as possible in the second sentence and write a short essay on them.

The responses far exceeded my expectations. To be sure, some of the undergraduates found the assignment unclear: I did not say how many errors they were supposed to find.[2] But many of the students put their hearts and souls

---

[1] My apologies to Prof. Henle, at whose expense I previously had a little fun on this matter. I used it again not because of any animosity I hold for him, but because I was familiar with it and, dealing with Euclid, it seemed appropriate for the start of my course.

[2] Fortunately, I did give instructions on spacing, font, and font size! Perhaps it is the way education courses are taught, but education majors expect everything to

into the exercise, some even finding fault with the first sentence of Henle's quote.

Henle's full quote contains two types of errors— those which everyone can agree are errors, and those I do not consider to be errors. The *bona fide* errors, in decreasing order of obviousness, are these: the date, the number of postulates, the extent of Euclid's coverage of mathematics, and Euclid's motivation in writing the *Elements*.

Different sources will present the student with different estimates of the dates of Euclid's birth and death, assuming they are bold enough to attempt such estimates. But they are consistent in saying he flourished around 300 B.C.[3] well after the 6th century B.C., which ran from 600 to 501 B.C., there being no year 0.

Some students suggested Henle may have got the date wrong because he was thinking of an earlier Euclid, namely Euclid of Megara, who was contemporary with Socrates and Plato. Indeed, mediæval scholars thought the two Euclids one and the same, and mention of Euclid of Megara in modern editions of Plato's dialogues is nowadays accompanied by a footnote explicitly stating that he of Megara is not *the* Euclid.[4] However, this explanation is incomplete: though he lived earlier than Euclid of Alexandria, Euclid of Megara still lived well after the 6th century B.C.

The explanation, if such is necessary, of Henle's placing of Euclid in the 6th century lies elsewhere, very likely in the 6th century itself. This was a century of great events— Solon reformed the laws of Athens; the religious leaders Buddha, Confucius, and Pythagoras were born; and western philosophy and theoretical mathematics had their origins in this century. That there might be more than two hundred years separating the first simple geometric propositions of Thales from a full blown textbook might not occur to someone living in our faster-paced times.

As to the number of postulates used by Euclid, Henle is correct that there are only five in the *Elements*. However, these are not the only assumptions Euclid based his development on. There were five additional axiomatic assertions he called "Common Notions", and he also used many definitions, some of which are axiomatic in character.[5] Moreover, Euclid made many implicit assumptions ranging from the easily overlooked (properties of betweenness and order) to the glaringly obvious (there is another dimension in solid geometry).

---

be spelled out for them, possibly because they are taught that they will have to do so at the levels they will be teaching.

[3] The referee informs me tht one eminent authority on Greek mathematics now dates Euclid at around 225 - 250 B.C.

[4] The conflation of the two Euclid's prompted me to exhibit in class the crown on the head of the astronomer Claudius Ptolemy in Raphæl's painting *The School of Athens*. Renaissance scholars mistakenly believed that Ptolemy, who lived in Alexandria under Roman rule, was one of the ptolemaic kings.

[5] E.g. I-17 asserts a diameter divides a circle in half; and V-4 is more-or-less the famous Axiom of Archimedes. (Cf. page 60, for more on this latter axiom.)

All students caught the incorrect date and most, if not all, were aware that Euclid relied on more than the 5 postulates. Some went on to explain the distinction between the notion of a postulate and that of an axiom,[6] a philosophical quibble of no mathematical significance, but a nice point to raise nevertheless. One or two objected that it was absurd to even imagine that all of mathematics could be derived from a mere 5 postulate's. This is either shallow and false or deep and true. In hindsight I realise I should have done two things in response to this. First, I should have introduced the class to Lewis Carroll's "What the Tortoise said to Achilles", which can be found in volume 4 of James R. Newman's *The World of Mathematics* cited in the Bibliography, below. Second, I should have given some example of amazing complexity generated by simple rules. Visuals go over well and, fractals being currently fashionable, a Julia set would have done nicely.

Moving along, we come to the question of Euclid's coverage. Did he really derive "most of the mathematical results known at the time"? The correct answer is, "Of course not". Euclid's *Elements* is a work on geometry, with some number theory thrown in. Proclus, antiquity's most authoritative commentator on Euclid, cites among Euclid's other works *Optics, Catoptics,* and *Elements of Music*— all considered mathematics in those days. None of the topics of these works is even hinted at in the *Elements,* which work also contains no references to conic sections (the study of which had been begun earlier by Menæchmus in Athens) or to such curves as the quadratrix or the conchoid which had been invented to solve the "three construction problems of antiquity". To quote Proclus:

> ... we should especially admire him for the work on the elements of geometry because of its arrangement and the choice of theorems and problems that are worked out for the instruction of beginners. He did not bring in everything he could have collected, but only what could serve as an introduction.[7]

In short, the *Elements* was not just a textbook, but it was an *introductory* textbook. There was no attempt at completeness[8].

---

[6] According to Proclus, a proposition is an axiom if it is known to the learner and credible in itself. If the proposition is not self-evident, but the student concedes it to his teacher, it is an hypothesis. If, finally, a proposition is unknown but accepted by the student as true without conceding it, the proposition is a postulate. He says, "axioms take for granted things that are immediately evident to our knowledge and easily grasped by our untaught understanding, whereas in a postulate we ask leave to assume something that can easily be brought about or devised, not requiring any labor of thought for its acceptance nor any complex construction".

[7] This is from page 57 of *A Commentary on the First Book of Euclid's Elements* by Proclus. Full bibliographic details are given in the Bibliography in the section on elementary mathematics.

[8] I used David Burton's textbook for the course. (Cf. the Bibliography for full bibliographic details.) On page 147 of the sixth edition we read, "Euclid tried to

This last remark brings us to the question of intent. What was Euclid's purpose in writing the *Elements*? Henle's appraisal that Euclid wrote the *Elements* as a result of his reflexion on the nature of the subject is not that implausible to one familiar with the development of set theory at the end of the 19th and beginning of the 20th centuries, particularly if one's knowledge of Greek mathematical history is a little fuzzy. Set theory began without restraints. Richard Dedekind, for example, proved the existence of an infinite set by referring to the set of his possible thoughts. This set is infinite because, given any thought S_0, there is also the thought S_1 that he is having thought S_0, the thought S_2 that he is having thought S_1, etc. Dedekind based the arithmetic of the real numbers on set theory, geometry was already based on the system of real numbers, and analysis (i.e., the Calculus) was in the process of being "arithmetised". Thus, all of mathematics was being based on set theory. Then Bertrand Russell asked the question about the set of all sets that were not elements of themselves:

$$R = \{x | x \notin x\}.$$

Is $R \in R$? If it is, then it isn't; and if it isn't, then it is.

The problem with set theory is that the naïve notion of set is vague. People mixed together properties of finite sets, the notion of property itself, and properties of the collection of subsets of a given unproblematic set. With hindsight we would expect contradictions to arise. Eventually Ernst Zermelo produced some axioms for set theory and even isolated a single clear notion of set for which his axioms were valid. There having been no contradictions in set theory since, it is a commonplace that Zermelo's axiomatisation of set theory was the reflexion and re-organisation[9] Henle suggested Euclid carried out— in Euclid's case presumably in response to the discovery of irrational numbers.

Henle did not precede his quoted remark with a reference to the irrationals, but it is the only event in Greek mathematics that could compel mathematicians to "pause and reflect", so I think it safe to take Henle's remark as asserting Euclid's axiomatisation was a response to the existence of these numbers. And this, unfortunately, ceases to be very plausible if one pays closer attention to dates. Irrationals were probably discovered in the 5th century B.C. and Eudoxus worked out an acceptable theory of proportions replacing the

---

build the whole edifice of Greek geometrical knowledge, amassed since the time of Thales, on five postulates of a specifically geometric nature and five axioms that were meant to hold for all mathematics; the latter he called common notions". It is enough to make one cry.

[9] Zermelo's axiomatisation was credited by David Hilbert with having saved set theory from inconsistency and such was Hilbert's authority that it is now common knowledge that Zermelo saved the day with his axiomatisation. That this was never his purpose is convincingly demonstrated in Gregory H. Moore, *Zermelo's Axiom of Choice; Its Origins, Development, and Influence*, Springer-Verlag, New York, 1982.

Pythagorean reliance on rational proportions in the 4th century. Euclid did a great deal of organising in the *Elements,* but it was not the necessity-driven response suggested by Henle, (or, my reading of him).[10]

So what was the motivation behind Euclid's work? The best source we have on this matter is the commentary on Book I of the *Elements* by Proclus in the 5th century A.D. According to Proclus, Euclid "thought the goal of the *Elements* as a whole to be the construction of the so-called Platonic figures" in Book XIII.[11] Actually, he finds the book to serve two purposes:

> If now anyone should ask what the aim of this treatise is, I should reply by distinguishing betweeen its purpose as judged by the matters investigated and its purpose with reference to the learner. Looking at its subject-matter, we assert that the whole of the geometer's discourse is obviously concerned with the cosmic figures. It starts from the simple figures and ends with the complexities involved in the structure of the cosmic bodies, establishing each of the figures separately but showing for all of them how they are inscribed in the sphere and the ratios that they have with respect to one another. Hence some have thought it proper to interpret with reference to the cosmos the purposes of individual books and have inscribed above each of them the utility it has for a knowledge of the universe. Of the purpose of the work with reference to the student we shall say that it is to lay before him an elementary exposition... and a method of perfecting... his understanding for the whole of geometry... This, then, is its aim: both to furnish the learner with an introduction to the science as a whole and to present the construction of the several cosmic figures.

The five platonic or cosmic solids cited are the tetrahedron, cube, octahedron, icosahedron, and dodecahedron. The Pythagoreans knew the tetrahedron, cube, and dodecahedron, and saw cosmic significance in them, as did Plato who had learned of the remaining two from Theætetus. Plato's speculative explanation of the world, the *Timæus* assigned four of the solids to the four elements: the tetrahedron to fire, the cube to earth, the icosahedron to water, and the octahedron to air. Later, Aristotle associated the dodecahedron with the æther, the fifth element. Euclid devoted the last book of the *Elements* to the platonic solids, their construction and, the final result of the book, the proof that these are the only regular solids. A neo-Platonist like Proclus would see great significance in this result and would indeed find it plausible that the presentation of the platonic solids could have been Euclid's goal[12]. Modern commentators don't find this so. In an excerpt from his trans-

---

[10] Maybe I am quibbling a bit? To quote the referee: "Perhaps Euclid didn't write the *Elements* directly in response to irrationals, but it certainly reflects a Greek response. And, historically, isn't that more important?"

[11] Op.cit., p. 57.

[12] Time permitting, some discussion of the Pythagorean-Platonic philosophy would be nice. I restricted myself to showing a picture of Kepler's infamous cosmological

lation of Proclus' commentary included by Drabkin and Cohen in *A Source Book in Greek Science,* G. Friedlein states simply, "One is hardly justified in speaking of this as the goal of the whole work".

A more modern historian, Dirk Struik says in his *Concise History of Mathematics,*

> What was Euclid's purpose in writing the *Elements*? We may assume with some confidence that he wanted to bring together into one text three great discoveries of the recent past: Eudoxus' theory of proportions, Theætetus' theory of irrationals, and the theory of the five regular bodies which occupied an outstanding place in Plato's cosmology. These three were all typically Greek achievements.

So Struik considers the *Elements* to be a sort of survey of recent research in a textbook for beginners.

From my student days I have a vague memory of a discussion between two Math Education faculty members about Euclid's *Elements* being not a textbook on geometry so much as one on geometric constructions. Specifically, it is a sort of manual on ruler and compass constructions. The opening results showing how to copy a line segment are explained as being necessary because the obvious trick of measuring the line segment with a compass and then positioning one of the feet of the compass at the point you want to copy the segment to could not be used with the collapsible compasses[13] of Euclid's day. The restriction to figures constructible by ruler and compass explains why conic sections, the quadratrix, and the conchoid are missing from the *Elements*. It would also explain why, in exhausting the circle, one continually doubles the number of sides of the required inscribed polygons: given an inscribed regular polygon of n sides, it is easy to further inscribe the regular $2n$-gon by ruler and compass construction, but how would one go about adding one side to construct the regular $(n + 1)$-gon? Indeed, this cannot in general be done.

The restriction of Euclid's treatment to figures and shapes constructible by ruler and compass is readily explained by the Platonic dictum that plane geometers restrict themselves to these tools. Demonstrating numerous constructions need not have been a goal in itself, but, like modern rigour, the rules of the game.

One thing is clear about Euclid's purpose in writing the *Elements*: he wanted to write a textbook for the instruction of beginners. And, while it is

---

representation of the solar system as a set of concentric spheres and inscribed regular polyhedra. As for mathematics, I used the platonic solid as an excuse to introduce Euler's formula relating the numbers of faces, edges, and vertices of a polyhedron and its application to classifying the regular ones. In Chapter 3, section 5, below, I use them for a different end.

[13] I have not done my homework. One of the referees made the remark, "Bell, isn't it?", indicating that I had too quickly accepted as fact an unsubstantiated conjecture by Eric Temple Bell.

clear he organised the material well, he cannot be said to have attempted to organise *all* of mathematical practice and derive most of it from his postulates.

Two errors uncovered by the students stretched things somewhat: did all of mathematical activity come to a complete stop in the "pause and reflect" process,— or, were some students taking an idiomatic "pause and reflect" intended to mean "reflect" a bit too literally? And: can Euclid be credited with deriving results if they were already known?

The first of these reputed errors can be dismissed out of hand. The second, however, is a bit puzzling, especially since a number of students misconstrued Henle as assigning priority to Euclid. Could it be that American education majors do not understand the process of derivation in mathematics? I have toyed with the notion that, on an ordinary reading of the word "derived", Henle's remark that Euclid derived most of the results known in his day from his postulates could be construed as saying that Euclid discovered the results. But I just cannot make myself believe it. Derivations are proofs and "deriving" means proving. To say that Euclid derived the results from his postulates says that Euclid showed that the results followed from his postulates, and it says no more; in particular, it in no way says the results (or even their proofs) originated with Euclid.

There was one more surprise some students had in store for me: Euclid was not a man, but a committee. This was not the students' fault. He, she, or they (I forget already) obviously came across this startling revelation in research-ing the problem. The *Elements* survives in 15 books, the last two of which are definitely not his and only the 13 canonical books are readily available. That these books are the work of a single author has been accepted for cen-turies. Proclus, who had access to many documents no longer available, refers to Euclid as a man and not as a committee. Nonetheless, some philologists have suggested multiple authors on the basis of linguistic analysis. Work by anonymous committee is not unknown in mathematics. In the twentieth cen-tury, a group of French mathematicians published a series of textbooks under the name Nicolas Bourbaki, which they had borrowed from an obscure Greek general. And, of course, the early Pythagoreans credited all their results to Pythagoras. These situations are not completely parallel: the composition of Bourbaki was an open secret, and the cult nature of the Pythagoreans widely known. Were Euclid a committee or the head of a cult, I would imagine some commentator would have mentioned it. Perhaps, however, we can reconcile the linguists with those who believe Euclid to have been one man by pointing to the German practice of the Professor having his lecture notes written up by his students after he has lectured?

2 About This Book

This book attempts to partially fill two gaps I find in the standard textbooks on the History of Mathematics. One is to provide the students with material that could encourage more critical thinking. General textbooks, attempting to cover three thousand or so years of mathematical history, must necessarily oversimplify just about everything, which practice can scarcely promote a critical approach to the subject. For this, I think a little narrow but deeper coverage of a few select topics is called for.

My second aim was to include the proofs of some results of importance one way or another for the history of mathematics that are neglected in the modern curriculum. The most obvious of these is the oft-cited necessity of introducing complex numbers in applying the algebraic solution of cubic equations. This solution, though it is now relegated to courses in the History of Mathematics, was a major occurrence in our history. It was the first substantial piece of mathematics in Europe that was not a mere extension of what the Greeks had done and thus signified the coming of age of European mathematics. The fact that the solution, in the case of three distinct real roots to a cubic, necessarily involved complex numbers both made inevitable the acceptance and study of these numbers and provided a stimulus for the development of numerical approximation methods. One should take a closer look at this solution.

Thus, my overall purpose in writing this book is twofold— to provide the teacher or student with some material that illustrates the importance of approaching history with a critical eye and to present the same with some proofs that are missing from the standard history texts.

In addition to this, of course, is the desire to produce a work that is not too boring. Thus, in a couple of chapters, I have presented the material as it unfolded to me. (In my discussion of Thomas Bradwardine in Chapter 3 I have even included a false start or two.) I would hope this would demonstrate to the student who is inclined to extract a term paper from a single source— as did one of my students did— what he is missing: the thrill of the hunt, the diversity of perspectives as the secondary and ternary authors each find something different to glean from the primary, interesting ancillary information and alternate paths to follow (as in Chapter 7, where my cursory interest in Horner's Method led me to Descartes' Rule and De Gua's Theorem), and an actual yearning for and true appreciation of primary sources.

I hope the final result will hold some appeal for students in a History of Mathematics course as well as for their teachers. And, although it may get bogged down a bit in some mathematical detail, I think it overall a good read that might also prove entertaining to a broader mathematical public. So, for better or worse, I unleash it on the mathematical public as is, as they say: warts and all.

Chapter 2 begins with a prefatory essay discussing many of the ways in which sources may be unreliable. This is followed by an annotated bibliography. Sometimes, but not always, the annotations rise to the occasion with critical comments.[14]

---

[14] It is standard practice in teaching the History of Mathematics for the instructor to hand out an annotated bibliography at the beginning of the course. But for

Chapter 3 is the strangest of the chapters in this book. It may serve to remind one that the nature of the real numbers was only finally settled in the 19th century. It begins with Pythagoras and all numbers being assumed rational and ends with Bradwardine and his proofs that the geometric line is not a discrete collection of points. The first proofs and comments offered on them in this chapter are solid enough; Bradwardine's proofs are outwardly nonsense, but there is something appealing in them and I attempt to find some intuition behind them. The critical mathematical reader will undoubtedly regard my attempt as a failure, but with a little luck he will have caught the fever and will try his own hand at it; the critical historical reader will probably merely shake his head in disbelief.

Chapters 4 to 7 are far more traditional. Chapter 4 discusses the construction problems of antiquity, and includes the proof that the angle cannot be trisected nor the cube duplicated by ruler and compass alone. The proof is quite elementary and ought to be given in the standard History of Mathematics course. I do, however, go well beyond what is essential for these proofs. I find the story rather interesting and hope the reader will criticise me for not having gone far enough rather than for having gone too far.

Chapter 5 concerns a Chinese word problem that piqued my interest. Ostensibly it is mainly about trying to reproduce the reasoning behind the original solution, but the account of the various partial representations of the problem in the literature provides a good example for the student of the need for consulting multiple sources when the primary source is unavailable to get a complete picture. I note that the question of reconstructing the probable solution to a problem can also profitably be discussed by reference to Plimpton 322 (a lot of Pythagorean triples or a table of secants), the Ishango bone (a tally stick or an "abacus" as one enthusiast described it), and the various explanations of the Egyptian value for π.

Chapter 6 discusses the cubic equation. It includes, as do all history textbooks these days, the derivation of the solution and examples of its application to illustrate the various possibilities. The heart of the chapter, however, is the proof that the algebraic solution uses complex numbers whenever the cubic equation has three distinct real solutions. I should say "proofs" rather than "proof". The first proof given is the first one to occur to me and was the first one I presented in class. It has, in addition to the very pretty picture on page 153, the advantage that all references to the Calculus can be stripped from it and it is, thus, completely elementary. The second proof is probably the easiest proof to follow for one who knows a little Calculus. I give a few other proofs and discuss some computational matters as well.

Chapter 7 is chiefly concerned with Horner's Method, a subject that usually merits only a line or two in the history texts, something along the lines of, "The Chinese made many discoveries before the Europeans. Horner's Method

_____

some editing and the addition of a few items, Chapter 2 is the one I handed out to my students.

is one of these." Indeed, this is roughly what I said in my course. It was only after my course was over and I was extending the notes I had passed out that I looked into Horner's Method, Horner's original paper, and the account of this paper given by Julian Lowell Coolidge in *The Mathematics of Great Amateurs*[15] that I realised that the standard account is oversimplified and even misleading. I discuss this in quite some detail before veering off into the tangential subjects of Descartes' Rule of Signs and something I call, for lack of a good name, De Gua's Theorem.

From discussion with others who have taught the History of Mathematics, I know that it is not all dead seriousness. One teacher would dress up for class as Archimedes or Newton... I am far too inhibited to attempt such a thing, but I would consider showing the occasional video[16]. And I do collect mathematicians on stamps and have written some high poetry— well, limericks— on the subject. I include this material in the closing Chapter 8, along with a couple of other historically interesting poems that may not be easily accessible.

Finally, I note that a short appendix outlines a few small projects, the likes of which could possibly serve as replacements for the usual term papers.

One more point— most students taking the History of Mathematics courses in the United States are education majors, and the most advanced mathematics they will get to teach is the Calculus. Therefore, I have deliberately tried not to go beyond the Calculus in this book and, whenever possible, have included Calculus-free proofs. This, of course, is not always possible.

---

[15] Cf. the Annotated Bibliography for full bibliographic details.
[16] I saw some a couple of decades ago produced, I believe, by the Open University in London and thought them quite good.

2

Annotated Bibliography

1 General Remarks

Historians distinguish between primary and secondary or even ternary sources. A primary source for, say, a biography would be a birth or death record, personal letters, handwritten drafts of papers by the subject of the biography, or even a published paper by the subject. A secondary source could be a biography written by someone who had examined the primary sources, or a non-photographic copy of a primary source. Ternary sources are things pieced together from secondary sources— encyclopædia or other survey articles, term papers, etc.[1] The historian's preference is for primary sources. The further removed from the primary, the less reliable the source: errors are made and propagated in copying; editing and summarising can omit relevant details, and replace facts by interpretations; and speculation becomes established fact even though there is no evidence supporting the "fact".[2]

1.1 Exercise. Go to the library and look up the French astronomer Camille Flammarion in as many reference works as you can find. How many different birthdays does he have? How many days did he die? If you have access to *World Who's Who in Science,* look up Carl Auer von Welsbach under "Auer" and "von Welsbach". What August day of 1929 did he die on?

---

[1] As one of the referees points out, the book before you is a good example of a ternary source.

[2] G.A. Miller's "An eleventh lesson in the history of mathematics", *Mathematics Magazine* 21 (1947), pp. 48 - 55, reports that Moritz Cantor's groundbreaking German language history of mathematics was eventually supplied with a list of 3000 errors, many of which were carried over to Florian Cajori's American work on the subject before the corrections were incorporated into a second edition of Cantor.

Answers to the Flammarion question will depend on your library. I found 3 birthdates and 4 death dates.[3] As for Karl Auer, the *World Who's Who in Science* had him die twice— on the 4th and the 8th. Most sources I checked let him rest in peace after his demise on the 4th. In my researches I also discovered that Max Planck died three nights in a row, but, unlike the case with von Welsbach, this information came from 3 different sources. I suspect there is more than mere laziness involved when general reference works only list the years of birth and death. However, even this is no guarantee of correctness: according to my research, the 20th century French pioneer of aviation Clément Ader died in 1923, again in 1925, and finally in 1926.

1.2 Exercise. Go to your favourite encyclopædia and read the article on Napoleon Bonaparte. What is Napoleon's Theorem?

In a general work such as an encyclopædia, the relevant facts about Napoleon are military and political. That he was fond of mathematics and discovered a theorem of his own is not a relevant detail. Indeed, for the history of science his importance is as a patron of the art and not as a a contributor. For a course on the history of mathematics, however, the existence of Napoleon's Theorem becomes relevant, if hardly central.

Translations, by their very nature, are interpretations. Sometimes in translating mathematics, a double translation is made: from natural language to natural langauge and then into mathematical language. That the original was not written in mathematical language could be a significant detail that is omitted. Consider only the difference in impressions that would be made by two translations of al-Khwarezmi's algebra book, one faithfully symbol-less in which even the number names are written out (i.e., "two" instead of "2") and one in which modern symbolism is supplied for numbers, quantities, and arithmetic operations. The former translation will be very heavy going and it will require great concentration to wade through the problems. You will be impressed by al-Khwarezmi's mental powers, but not by his mathematics as it will be hard to survey it all in your mind. The second translation will be easy going and you shouldn't be too impressed unless you mistakenly believe, from the fact that the word "algebra" derived from the Arabic title of his book, that the symbolic approach originated here as well.

The first type of translation referred to is the next best thing to the primary source. It accurately translates the contents and allows the reader to interpret them. The second type accurately portrays the problems treated, as well as the abstract principles behind the methods, possibly more as a concession to readability than a conscious attempt at analysis, but in doing so it does not accurately portray the actual practice and may lead one to overestimate the original author's level of understanding. Insofar as a small shift in one's

---

[3] I only found them in 4 different combinations. However, through clever footnoting and the choice of different references for the birth and death dates, I can justify $3 \times 4 = 12$ pairs!

perspective can signify a major breakthrough, such a translation can be a significant historical distortion.

It is important in reading a translation to take the translator's goal into account, as revealed by the following quotation from Samuel de Fermat (son of *the* Fermat) in his preface to a 1670 edition of Diophantus:

> Bombelli in his *Algebra* was not acting as a translator for Diophantus, since he mixed his own problems with those of the Greek author; neither was Viète, who, as he was opening up new roads for algebra, was concerned with bringing his own inventions into the limelight rather than with serving as a torch-bearer for those of Diophantus. Thus it took Xylander's unremitting labours and Bachet's admirable acumen to supply us with the translation and interpretation of Diophantus's great work.[4]

And, of course, there is always the possibility of a simple mistranslation. My favourite example was reported by the German mathematical educator Herbert Meschkowski.[5] The 19th century constructivist mathematician Leopold Kronecker, in criticising abstract mathematical concepts, declared, "Die ganzen Zahlen hat der liebe Gott gemacht. Alles andere ist Menschenwerk." This translates as "The Good Lord made the whole numbers. Everything else is manmade", though something like "God created the integers; all the rest is man's work" is a bit more common. The famous theologian/mystery novelist Dorothy Sayers quoted this in one of her novels, which was subsequently translated into German. Kronecker's remark was rendered as "Gott hat die Integralen erschaffen. Alles andere ist Menschenwerk", or "God has created the integrals. All the rest is the work of man"!

Even more basic than translation is transliteration. When the matchup between alphabets is not exact, one must approximate. There is, for example, no equivalent to the letter "h" in Russian, whence the Cyrillic letter most closely resembling the Latin "g" is used in its stead. If a Russian paper mentioning the famous German mathematician David Hilbert is translated into English by a nonmathematician, Hilbert's name will be rendered "Gilbert", which, being a perfectly acceptable English name, may not immediately be recognised by the reader as "Hilbert". Moreover, the outcome will depend on the nationality of the translator. Thus the Russian mathematician Chebyshev's name can also be found written as Tchebichev (French) and Tschebyschew (German). Even with a fixed language, transliteration is far from unique, as schemes for transliteration change over time as the reader will see when we get to the chapter on the Chinese word problem. But we are digressing.

We were discussing why primary sources are preferred and some of the ways references distant from the source can fail to be reliable. I mentioned

---

[4] Quoted in André Weil, *Number Theory; An Approach Through History, From Hammurapi to Legendre*, Birkhäuser, Boston, 1984, p. 32.

[5] *Mathematik und Realität, Vorträge und Aufsätze*, Bibliographisches Institut, Mannheim, 1979, p. 67.

above that summaries can be misleading and can replace facts by interpretation. A good example is the work of Diophantus, whose *Arithmetica* was a milestone in Greek mathematics. Diophantus essentially studied the problem of finding positive rational solutions to polynomial equations. He introduced some symbolism, but not enough to make his reasoning easily accessible to the modern reader. Thus one can find summary assessments— most damningly expressed in Eric Temple Bell's *Development of Mathematics*,[6]— to the effect that Diophantus is full of clever tricks, but possesses no general methods. Those who read Diophantus 40 years after Bell voiced a different opinion: Diophantus used techniques now familiar in algebraic geometry, but they are hidden by the opacity of his notation. The facts that Diophantus solved this problem by doing this, that one by doing that, etc., were replaced in Bell's case by the interpretation that Diophantus had no method, and in the more modern case, by the diametrically opposed interpretation that he had a method but not the language to describe it.

Finally, as to speculation becoming established fact, probably the quintessential example concerns the Egyptian rope stretchers. It is, I believe, an established fact that the ancient Egyptians used rope stretchers in surveying. It is definitely an established fact that the Pythagorean Theorem and Pythagorean triples like 3, 4, 5 were known to many ancient cultures. Putting 2 and 2 together, the German historian Moritz Cantor speculated that the rope stretchers used knotted ropes giving lengths 3, 4, and 5 units to determine right angles. To cite Bartel van der Wærden,[7]

> ...How frequently it happens that books on the history of mathematics copy their assertions uncritically from other books, without consulting the sources... In 90% of all the books, one finds the statement that the Egyptians knew the right triangle of sides 3, 4, and 5, and that they used it for laying out right triangles. How much value has this statement? None!

Cantor's conjecture is an interesting possibility, but it is pure speculation, not backed up by any evidence that the Egyptians had any knowledge of the Pythagorean Theorem at all. Van der Wærden continues

> To avoid such errors, I have checked all the conclusions which I found in modern writers. This is not as difficult as might appear... For reliable translations are obtainable of nearly all texts...
> Not only is it more instructive to read the classical authors themselves (in translation if necessary), rather than modern digests, it also gives much greater enjoyment.

Van der Wærden is not alone in his exhortation to read the classics, but "obtainable" is not the same as "readily available" and one will have to rely on

---

[6] McGraw-Hill, New York, 1940

[7] *Science Awakening*, 2nd ed., Oxford University Press, New York, 1961, p. 6.

"digests", general reference works, and other secondary and ternary sources for information. Be aware, however, that the author's word is not gospel. One should check if possible the background of the author: does he or she have the necessary mathematical background to understand the material; what sources did he/she consult; and, does the author have his/her own axe to grind?

Modern history of mathematics began to be written in the 19th century by German mathematicians, and several histories were written by American mathematicians in the early 20th century. And today much of the history of mathematics is still written by mathematicians. Professional historians traditionally ignored the hard technical subjects simply because they lacked the understanding of the material involved. In the last several decades, however, a class of professional historians of science trained in history departments has arisen and some of them are writing on the history of mathematics. The two types of writers tend to make complementary mistakes— or, at least, be judged by each other as having made these mistakes.

Some interdisciplinary errors do not amount to much. These can occur when an author is making a minor point and adds some rhetorical flourish without thinking too deeply about it. We saw this with Henle's comment on Euclid in the introduction. I don't know how common it is in print, but its been my experience that historical remarks made by mathematicians in the classroom are often simply factually incorrect. These same people who won't accept a mathematical result from their teachers without proof will accept their mentors' anecdotes as historical facts. Historians' mistakes at this level are of a different nature. Two benign examples come to mind. Joseph Dauben, in a paper[8] on the Chinese approach to the Pythagorean Theorem, compares the Chinese and Greek approaches with the remark that

> ...whereas the Chinese demonstration of the right-triangle theorem involves a rearrangement of areas to show their equivalence, EUCLID's famous proof of the Pythagorean Theorem, Proposition I,47, does not rely on a simple shuffling of areas, moving a to b and c to d, but instead depends upon an elegant argument requiring a careful sequence of theorems about similar triangles and equivalent areas.

The mathematical error here is the use of the word "similar", the whole point behind Euclid's complex proof having been the avoidance of similarity which depends on the more advanced theory of proportion only introduced later in Book V of the *Elements*.[9]

---

[8] Joseph Dauben, "The 'Pythagorean theorem' and Chinese Mathematics. Liu Hui's Commentary on the Gou-Gu Theorem in Chapter Nine of the *Jin Zhang Suan Shu*", in: S.S. Demidov, M. Folkerts, D.E. Rowe, and C.J. Scriba, eds., *Amphora; Festschrift für Hans Wussing zu seinem 65. Geburtstag*, Birkhäuser-Verlag, Basel, 1992.

[9] Cf. the chapter on the foundations of geometry for a fuller discussion of this point. Incidentally, the use of the word "equivalent" instead of "equal" could also

Another example of an historian making an inconsequential mathematical error is afforded us by Ivor Grattan-Guinness, but concerns more advanced mathematics. When he discovered some correspondence between Kurt Gödel and Ernst Zermelo concerning the former's famous Incompleteness Theorem, he published it along with some commentary[10]. One comment was that Gödel said his proof was nonconstructive. Now anyone who has read Gödel's original paper can see that the proof is eminently constructive and would doubt that Gödel would say such a thing. And, indeed, he didn't. What Gödel actually wrote to Zermelo was that an alternate proof related to Zermelo's initial criticism was— unlike his published proof— nonconstructive. Grattan-Guiness had simply mistranslated and thereby stated something that was mathematically incorrect.

Occasionally, the disagreement between historian and mathematician can be serious. The most famous example concerns the term "geometric algebra", coined by the Danish mathematician Hieronymus Georg Zeuthen in the 1880s to describe the mathematics in one of the books of the *Elements*. One historian saw in this phrase a violation of basic principles of historiography and proposed its banishment. His suggestion drew a heated response that makes for entertaining reading.[11]

---

be considered an error by mathematicians. For, areas being numbers they are either equal or unequal, not equivalent.

[10] I. Grattan-Guinness, "In memoriam Kurt Gödel: his 1931 correspondence with Zermelo on his incompletability theorem", *Historia Mathematica* 6 (1979), pp. 294 - 304.

[11] The initial paper and all its responses appeared in the *Archive for the History of the Exact Sciences*. The first, somewhat polemical paper, "On the need to rewrite the history of Greek mathematics" (vol. 15 (1975/76), pp. 67 - 114) was by Sabetai Unguru of the Department of the History of Science at the University of Oklahoma and about whom I know only this controversy. The respondents were Bartel van der Wærden ("Defence of a 'shocking' point of view", vol. 15 (1975), pp. 199 - 210), Hans Freudenthal ("What is algebra and what has been its history?", vol. 16 (1976/77), pp. 189 - 200), and André Weil ("Who betrayed Euclid", vol. 19 (1978), pp. 91 - 93), big guns all. The Dutch mathematician van der Wærden is particularly famous in the history of science for his book *Science Awakening,* which I quoted from earlier. He also authored the classic textbook on modern algebra, as well as other books on the history of early mathematics. Hans Freudenthal, another Dutch mathematician, was a topologist and a colourful character who didn't mince words in the various disputes he participated in during his life. As to the French André Weil, he was one of the leading mathematicians of the latter half of the 20th century. Regarding his historical qualifications, I cited his history of number theory earlier. Unguru did not wither under the massive assault, but wrote a defence which appeared in a different journal: "History of ancient mathematics; some reflections on the state of the art", *Isis* 20 (1979), pp. 555 - 565. Perhaps the editors of the *Archive* had had enough. Both sides had valid points and the dispute was more a clash of perspectives than anyone making major errors. Unguru's *Isis* paper is worth a read. It may be opaque in spots,

On the subject of the writer's motives, there is always the problem of the writer's ethnic, religious, racial, gender, or even personal pride getting in the way of his or her judgement. The result is overstatement.

In 1992, I picked up a paperback entitled *The Miracle of Islamic Science*[12] by Dr. K. Ajram. As sources on Islamic science are not all that plentiful, I was delighted— until I started reading. Ajram was not content to enumerate Islamic accomplishments, but had to ignore earlier Greek contributions and claim priority for Islam. Amidst a list of the "sciences originated by the muslims" he includes trigonometry, apparently ignorant of Ptolemy, whose work on astronomy beginning with the subject is today known by the name given it by the Arabic astronomers who valued it highly. His attempt to denigrate Copernicus by assigning priority to earlier Islamic astronomers simply misses the point of Copernicus's accomplishments, which was not merely to place the sun in the centre of the solar system— which was in fact already done by Aristarchus centuries before Islam or Islamic science existed, a fact curiously unmentioned by Ajram. Very likely most of his factual data concerning Islamic science is correct, but his enthusiasm makes his work appear so amateurish one cannot be blamed for placing his work in the "unreliable" stack.[13]

Probably the most extreme example of advocacy directing history is the Afrocentrist movement, an attempt to declare black Africa to be the source of all Western Culture. The movement has apparently boosted the morale of Africans embarrassed at their having lagged behind the great civilisations of Europe and Asia. I have not read the works of the Afrocentrists, but if one may judge from the responses to it,[14][15] emotions must run high. The Afrocentrists have low standards of proof (Example: Socrates was black for i. he was not from Athens, and ii. he had a broad nose.) and any criticism is apparently met with a charge of racism. (Example: the great historian of ancient astronomy, Otto Neugebauer described Egyptian astronomy as "primitive" and had better things to say about Babylonian astronomy. The reason for this was declared by one prominent Afrocentrist to be out and out racial prejudice against black

---

and not as much fun to read as the attacks, but it does offer a good discussion of some of the pitfalls in interpreting history.

An even earlier clash between historian and mathematician occurred in the pages of the *Archive* when Freudenthal pulled no punches in his response ("Did Cauchy plagiarize Bolzano?", 7 (1971), pp. 375 - 392) to a paper by Grattan-Guinness ("Bolzano, Cauchy, and the 'new analysis' of the early nineteenth century", 6 (1969/70), pp. 372 - 400).

[12] Knowledge House Publishers, Cedar Rapids, 1992

[13] The referee points out that "the best example of distortion due to nationalist advocacy is early Indian science". I have not looked into this.

[14] Robert Palter, "*Black Athena,* Afrocentrism and the History of Science," *History of Science* 31 (1993), pp. 227 - 287.

[15] Mary Lefkowitz, *Not Out of Africa; How Afrocentrism Became an Excuse to Teach Myth as History,* New Republic Books, New York, 1996.

Egyptians and preference for the white Babylonians. The fact of the greater sophistication and accuracy of the Babylonian practice is irrelevant.)

Let me close with a final comment on an author's agenda. He may be presenting a false picture of history because history is not the point he is trying to get across. Samuel de Fermat's remarks on Bombelli and Viète cited earlier are indications. These two authors had developed techniques the usefulness of which they wanted to demonstrate. Diophantus provided a stock of problems. Their goal was to show how their techniques could solve these problems and others, not to show how Diophantus solved them. In one of my own books, I wanted to discuss Galileo's confusions about infinity. This depended on two volume calculations which he did geometrically. I replaced these by simple applications of the Calculus on the grounds that my readers would be more familiar with the analytic method. The relevant point here was the shared value of the volumes and not how the result was arrived at, just as for Bombelli and Viète the relevant point would have been a convenient list of problems. These are not examples of bad history, because they are not history at all. Ignoring the context and taking them to be history would be the mistake here.

So there we have a discussion of some of the pitfalls in studying the history of mathematics. I hope I haven't convinced anyone that nothing one reads can be taken as true. This is certainly not the case. Even the most unreliable sources have more truth than fiction to them. The problem is to sort out which statements are indeed true. For this course, the best guarantee of the reliability of information is endorsement of the author by a trusted authority (e.g., your teacher). So without further ado, I present the following annotated bibliography.

2 General Reference Works

Encyclopædia Britannica

This is the most complete encyclopædia in the English language. It is very scholarly and generally reliable. However, it does not always include scientific information on scientifically marginal figures.

Although the edition number doesn't seem to change these days, new printings from year to year not only add new articles, but drop some on less popular subjects. It is available in every public library and also online.

Any university worthy of the name will also have the earlier 11th edition, called the "scholar's edition". Historians of science actually prefer the even earlier 9th edition, which is available in the libraries of the better universities. However, many of the science articles of the 9th edition were carried over into the 11th.

Enciclopedia Universal Ilustrada Europeo-Americana

The Spanish encyclopædia originally published in 70 volumes, with a 10 volume appendix, is supplemented each year.

I am in no position to judge its level of scholarship. However, I do note that it seems to have the broadest selection of biographies of any encyclopædia, including, for example, an English biologist I could find no information on anywhere else. In the older volumes especially, birth and death dates are unreliable. These are occasionally corrected in the later supplements.

Great Soviet Encyclopedia, 3rd Edition, MacMillan Inc., New York, 1972 - 1982.

Good source for information on Russian scientists. It is translated volume by volume, and entries are alphabetised in each volume, but not across volumes. Thus, one really needs the index volume or a knowledge of Russian to look things up in it. It is getting old and has been removed from the shelves of those few suburban libraries I used to find it in. Thus one needs a university library to consult it.

3 General Biography

J.C. Poggendorff, *Biographisch-literarisches Handwörterbuch zur Geschichte der exacten Wissenschaften*

This is the granddaddy of scientific biography. Published in the mid-19th century with continuing volumes published as late as 1926, the series received an American *Raubdruck*[16] edition in 1945 and is consequently available in some of the better universities. The entries are mostly short, of the *Who's Who* variety, but the coverage is extensive. Birth and death dates are often in error, occasionally corrected in later volumes.

Allen G. Debus, ed., *World Who's Who of Science; From Antiquity to the Present*

Published in 1968 by the producers of the *Who's Who* books, it contains concise *Who's Who* styled entries on approximately 30000 scientists. Debus is an historian of science and the articles were written by scholars under his direction. Nonetheless, there are numerous incorrect birth and death dates and coordination is lacking as some individuals are given multiple, non-cross-referenced entries under different names.

---

[16] That is, the copyright was turned over to an American publisher by the US Attorney General as one of the spoils of war.

The preface offers a nice explanation of the difficulties involved in creating a work of this kind and the errors that are inherent in such an undertaking.

I have found the book in some municipal libraries and not in some university libraries.

Charles Gillespie, ed., *Dictionary of Scientific Biography*, Charles Scribner's Sons, New York, 1970 - 1991.

This encyclopædia is the best first place to find information on individual scientists who died before 1972. It consists of 14 volumes of extensive biographical articles written by authorities in the relevant fields, plus a single volume supplement, and an index. Published over the years 1970 - 1980, it was augmented in 1991 by an additional 2 volumes covering those who died before 1981.

The *Dictionary of Scientific Biography* is extremely well researched and most reliable. As to the annoying question of birth and death dates, the only possible error I found is Charles Darwin's birthdate, which disagrees with all other references I've checked, including Darwin's autobiography. I suspect Darwin was in error and all the other sources relied on his memory...

The *Dictionary of Scientific Biography* is available in all university and most local libraries.

A *Biographical Dictionary of Mathematicians* has been culled from the *Dictionary of Scientific Biography* and may interest those who would like to have their own copy, but cannot afford the complete set.

Eric Temple Bell, *Men of Mathematics*, Simon and Schuster, New York, 1937.

First published in 1937, this book is still in print today. It is a popularisation, not a work of scholarship, and Bell gets important facts wrong. However, one does not read Bell for information, but for the sheer pleasure of his impassioned prose.

Julian Lowell Coolidge, *The Mathematics of Great Amateurs,* Oxford University Press, Oxford, 1949.

A Dover paperback edition appeared in 1963, and a new edition edited by Jeremy Gray was published by Oxford University Press in 1990. What makes this book unique are i) the choice of subjects and ii) the mathematical coverage. The subjects are people who were not primarily mathematicians— the philosophers Plato and Pascal, the artists Leonardo da Vinci and Albrecht Dürer, a politician, some aristocrats, a school teacher, and even a theologian. The coverage is unusual in that Coolidge discusses the mathematics of these great amateurs. In the two chapters I read carefully I found errors.

Isaac Asimov, *Asimov's Biographical Encyclopedia of Science and Technology*, Doubleday, New York, 1982.

> This is a one-volume biographical dictionary, not an encyclopædia, with entries chronologically organised.
> One historian expressed horror to me at Asimov's methodology. So he would be an acceptable source as a reference for a term paper, but his use in a thesis would be cause for rejection. The problem is that the task he set for himself is too broad for one man to perform without relying on references far removed from the primary sources.

This list could be endlessly multiplied. There are several small collections like Bell's of chapter-sized biographies of a few mathematicians, as well as several large collections like Asimov's of short entry biographies of numerous mathematicians and scientists. For the most part, one is better off sticking to the *Dictionary of Scientific Biography* or looking for a dedicated biography of the individual one is interested in. That said, I note that works like E.G.R. Taylor's *The Mathematical Practitioners of Tudor and Stuart England* (Cambridge University Press, 1954) and *The Mathematical Practitioners of Hanoverian England* (Cambridge University Press, 1966), with their 3500 mini-biographies and essays on mathematical practice other than pure mathematical research are good sources for understanding the types of uses mathematics was being put to in these periods.

4 General History of Mathematics

Florian Cajori, *History of Mathematics*, Macmillan and Company, New York, 1895.

—, *A History of Elementary Mathematics, with Hints on Methods of Teaching*, The Macmillan Company, New York, 1917.

—, *A History of Mathematical Notations*, 2 volumes, Open Court Publishing Company, Lasalle (Ill), 1928 - 29.

> The earliest of the American produced comprehensive histories of mathematics is Cajori's, which borrowed a lot from Moritz Cantor's monumental four volume work on the subject, including errors. Presumably most of these have been corrected through the subsequent editions. The current edition is a reprint of the 5th published by the American Mathematical Society.
> Cajori's history of elementary mathematics was largely culled from the larger book and is no longer in print.
> Cajori's history of mathematical notation is a cross between a reference work and a narrative. A paperback reprint by Dover Publishing Company exists.

David Eugene Smith, *History of Mathematics,* 2 volumes, 1923, 1925.

—, *A Source Book in Mathematics,,* 1929.

—, *Rara Arithmetica,* Ginn and Company, Boston, 1908.

All three books are in print in inexpensive Dover paperback editions.
The first of these was apparently intended as a textbook, or a history
for mathematics teachers as it has "topics for discussion" at the end
of each chapter. Most of these old histories do not have much actual
mathematics in them. The second book complements the first with a
collection of excerpts from classic works of mathematics.
Rara Arithmetica is a bibliographic work, describing a number of old
mathematics books, which is much more interesting than it sounds.

Eric Temple Bell, *Development of Mathematics,* McGraw-Hill, New York,
1940.

Bell is one of the most popular of American writers on mathematics
of the first half of the 20th century and his books are still in print.
There is nothing informational in this history to recommend it over
the others listed, but his style and prose beat all the rest hands down.

Dirk Struik, *A Concise History of Mathematics,* revised edition, Dover, New
York, 1967.

This is considered by some to be the finest short account of the history
of mathematics, and it very probably is. However, it is a bit too concise
and I think one benefits most in reading it for additional insight after
one is already familar with the history of mathematics.

Howard Eves, *An Introduction to the History of Mathematics,* Holt, Rinehart,
and Winston, New York, 1953.

Carl Boyer, *A History of Mathematics,* John Wiley and Sons, New York, 1968.

Both books have gone through several editions and, I believe, are still
in print. They were written specifically for the class room and included
genuine mathematical exercises. Eves peppers his book (at least, the
edition I read) with anecdotes that are most entertaining and reveal
the "human side" of mathematicians, but add nothing to one's understanding of the development of mathematics. Boyer is much more
serious. The first edition was aimed at college juniors and seniors in a
post-Sputnik age of higher mathematical expectations; if the current
edition has not been watered down, it should be accessible to some
seniors and to graduate students. Eves concentrates on elementary
mathematics, Boyer on calculus.
Both author's have written other books on the history of mathematics. Of particular interest are Boyer's separate histories of analytic
geometry and the calculus.

David M. Burton, *The History of Mathematics; An Introduction,* McGraw-Hill, New York, 1991.

Victor J. Katz, *A History of Mathematics; An Introduction,* Harper Collins, New York, 1993.

These appear to be the current textbooks of choice for the American market and are both quite good. A publisher's representative for McGraw-Hill informs me Burton's is the best-selling history of mathematics textbook on the market, a claim supported by the fact that, as I write, it has just come out in a 6th edition. Katz is currently in its second edition. One referee counters with, "regardless of sales, Katz is considered the standard textbook at its level by professionals".[17] Another finds Burton "systematically unreliable". I confess to having found a couple of howlers myself.

Both books have a lot of history, and a lot of mathematical exercises. Katz's book has more mathematics and more advanced mathematics than the other textbooks cited thus far.

Roger Cooke, *The History of Mathematics; A Brief Course,* Wiley Interscience, 1997.

I haven't seen this book, which is now in its second edition (2005). Cooke has excellent credentials in the history of mathematics and I would not hesitate in recommending his book sight unseen. The first edition was organised geographically or culturally— first the Egyptians, then Mesopotamians, then Greeks, etc. The second edition is organised by topic— number, space, algebra, etc. Both are reported strong on discussing the cultural background to mathematics.

Morris Kline, *Mathematical Thought from Ancient to Modern Times,* Oxford University Press, New York, 1972.

This is by far the best single-volume history of general mathematics in the English language that I have seen. It covers even advanced mathematical topics and 20th century mathematics. Kline consulted many primary sources and each chapter has its own bibliography.

5 History of Elementary Mathematics

Otto Neugebauer, *The Exact Sciences in Antiquity,* Princeton University Press, Princeton, 1952.

B.L. van der Wærden, *Science Awakening,* Oxford university Press, New York, 1961.

---

[17] The referee did not say whether these are professional historians, mathematicians, or teachers of the history of mathematics.

These are the classic works on mathematics and astronomy from the Egyptians through the Hellenistic (i.e. post-Alexander) period. Van der Wærden's book contains more mathematics and is especially recommended. It remains in print in a Dover paperback edition.

Lucas N.H. Bunt, Phillip S. Jones, and Jack D. Bedient, *The Historical Roots of Elementary Mathematics,* Prentice Hall, Englewood Cliffs (New Jersey), 1976.

This is a textbook on the subject written for a very general audience, presupposing only high school mathematics. It includes a reasonable number of exercises. A Dover reprint exists.

Asger Aaboe, *Episodes From the Early History of Mathematics,* Mathematical Association of America, 1964.

This slim volume intended for high school students includes expositions of some topics from Babylonian and Greek mathematics. A small number of exercises is included.

Richard Gillings, *Mathematics in the Time of the Pharoahs,* MIT Press, Cambridge (Mass), 1973

This and Gilling's later article on Egyptian mathematics published in the *Dictionary of Scientific Biography* offer the most complete treatments of the subject readily available. It is very readable and exists in an inexpensive Dover paperback edition.

Euclid, *The Elements*

Proclus, *A Commentary on the First Book of Euclid's Elements*, translated by Glenn Morrow, Princeton University Press, Princeton, 1970.

The three most accessible American editions of *The Elements* are Thomas Heath's translation, available in the unannotated *Great Books of the Western World* edition, an unannotated edition published by Green Lion Press, and a super-annotated version published in 3 paperback volumes from Dover. The Dover edition is the recommended version because of the annotations. If one doesn't need or want the annotations, the Green Lion Press edition is the typographically most beautiful of the three and repeats diagrams on successive pages for greater ease of reading. But be warned: Green Lion Press also published an abbreviated outline edition not including the proofs.

Proclus is an important historical document in the history of Greek mathematics for a variety of reasons. Proclus had access to many documents no longer available and is one of our most detailed sources of early Greek geometry. The work is a good example of the commentary that replaced original mathematical work in the later periods of Greek mathematical supremacy. And, of course, it has much to say about Euclid's *Elements*.

Howard Eves, *Great Moments in Mathematics (Before 1650)*, Mathematics Association of America, 1980.

> This is a book of short essays on various developments in mathematics up to the eve of the invention of the Calculus (which is covered in a companion volume). It includes historical and mathematical exposition as well as exercises. I find the treatments a bit superficial, but the exercises counter this somewhat.

6 Source Books

A source book is a collection of extracts from primary sources. The first of these, still in print, was Smith's mentioned earlier:

David Eugene Smith, *A Source Book in Mathematics*

> At a more popular level is the following classic collection.

James R. Newman, *The World of Mathematics*, Simon and Schuster, New York, 1956.

> This popular 4 volume set contains a wealth of material of historical interest. It is currently available in a paperback edition.

Ivor Thomas, *Selections Illustrating the History of Greek Mathematics, I; Thales to Euclid*, Harvard University Press, Cambridge (Mass), 1939.

—, *Selections Illustrating the History of Greek Mathematics, II; From Aristarchus to Pappus*, Harvard University Press, Cambridge (Mass), 1941.

> These small volumes from the Loeb Classical Library are presented with Greek and English versions on facing pages. There is not a lot, but the assortment of selections was judiciously made.

Morris R. Cohen and I.E. Drabkin, *A Source Book in Greek Science*, Harvard University Press, Cambridge (Mass), 1966.

Edward Grant, *A Source Book in Medieval Science*, Harvard University Press, Cambridge (Mass), 1974.

Dirk Struik, *A Source Book in Mathematics, 1200 - 1800*, Harvard University Press, Cambridge (Mass), 1969.

> In the 1960s and 1970s, Harvard University Press published a number of fine source books in the sciences. The three listed are those most useful for a general course on the history of mathematics. More advanced readings can be found in the specialised source books in analysis and mathematical logic. I believe these are out of print, but I would expect them to be available in any university library.

Ronald Calinger, *Classics of Mathematics,* Moore Publishing Company, Oak Park (Ill), 1982.

> For years this was the only general source book for mathematics to include twentieth century mathematics. The book is currently published by Prentice-Hall.

Douglas M. Campbell and John C. Higgins, *Mathematics; People, Problems, Results,* Wadsworth International, Belmont (Cal), 1984.

> This three volume set was intended to be an up-to-date replacement for Newman's *World of Mathematics.* It's extracts, however, are from secondary sources rather than from primary sources. Nonetheless it remains of interest.

John Fauvel and Jeremy Gray, *The History of Mathematics; A Reader,* McMillan Education, Ltd, London, 1987.

> This is currently published in the US by the Mathematical Association of America. It is probably the nicest of the source books. In addition to extracts from mathematical works, it includes extracts from historical works (e.g., comments on his interpretation of the Ishango bone by its discoverer, and extracts from the debate over Greek geometric algebra) and some cultural artefacts (e.g., Alexander Pope and William Blake on Newton).

Stephen, Hawking, *God Created the Integers; The Mathematical Breakthroughs that Changed History,* Running Press, Philadelphia, 2005.

> The blurb on the dust jacket and the title page announce this collection was edited with commentary by Stephen Hawking. More correctly stated, each author's works are preceded by an essay by the renowned physicist titled "His life and work"; explanatory footnotes and, in the case of Euclid's *Elements*, internal commentary are lifted without notice from the sources of the reproduced text. This does not make the book any less valuable, but if one doesn't bear this in mind one might think Hawking is making some statement about our conception of time when one reads the reference (which is actually in Thomas Heath's words) to papers published in 1901 and 1902 as having appeared "in the last few years". Aside from this, it is a fine collection, a judicious choice that includes some twentieth century mathematics with the works of Henri Lebesgue, Kurt Gödel, and Alan Turing.

Jean-Luc Chabert, ed., *A History of Algorithms, From the Pebble to the Microchip,* Springer-Verlag, Berlin, 1999.

> Originally published in French in 1994, this is a combination history and source book. I list it under source books rather than special historical topics because of the rich variety of the excerpts included and the breadth of the coverage, all areas of mathematics being subject to algorithmic pursuits.

7 Multiculturalism

George Gheverghese Joseph, *The Crest of the Peacock; Non-European Roots of Mathematics,* Penguin Books, London, 1992.

> A very good account of non-European mathematics which seems to be quite objective and free of overstatement.

Yoshio Mikami, *The Development of Mathematics in China and Japan,* 2nd. ed., Chelsea Publishing Company, New York, 1974.

Joseph Needham, *Science and Civilization in China, III; Mathematics and the Sciences of the Heavens and the Earth,* Cambridge University Press, Cambridge, 1959.

Lǐ Yan and Dù Shíràn, *Chinese Mathematics; A Concise History,* Oxford University Press, Oxford, 1987.

> Mikami's book was first published in German in 1913 and is divided into two parts on Chinese and Japanese mathematics, respectively. Needham's series of massive volumes on the history of science in China is the standard. The third volume covers mathematics, astronomy, geography, and geology and is not as technical as Mikami or the more recent book by Lǐ Yan and Dù Shíràn, for which Needham wrote the Foreword.
>
> Needham's book is still in print. The other two books are out of print.

David Eugene Smith and Yoshio Mikami, *A History of Japanese Mathematics*

> I haven't seen this book, but in the introductory note of his book on Chinese and Japanese mathematics, Mikami announces that the book was to be written at a more popular level. It is in print in 2 or 3 editions, including a paperback one by Dover.

Seyyed Hossein Nasr, *Science and Civilization in Islam,* Harvard University Press, Cambridge (Mass), 1968.

> Nasr borrowed the title from Needham, but his work is much shorter—only about 350 pages. It does not have much technical detail, and the chapter on mathematics is only some 20 odd pages long. The book is still in print in a paperback edition.

J.L. Berggren, *Episodes in the Mathematics of Medieval Islam,* Springer-Verlag, NY, 1986.

> This appears to be the best source on Islamic mathematics. It even includes exercises. The book is still in print.

8 Arithmetic

Louis Charles Karpinski, *The History of Arithmetic,* Rand McNally and Company, Chicago, 1925.

> This is the classic American study of numeration and computation by hand. It includes history of early number systems, the Hindu-Arabic numerals, and even a brief study of textbooks from Egypt to America and Canada. The book is out of print.

Karl Menninger, *Number Words and Number Symbols; A Cultural History of Mathematics,* MIT Press, Cambridge (Mass), 1969.

> This large volume covers the history of numeration and some aspects of the history of computation, e.g. calculation with an abacus. The book is currently in print by Dover.

9 Geometry

Adrien Marie Legendre, *Geometry*

> One of the earliest rivals to Euclid (1794), this book in English translation was the basis for geometry instruction in United States in the 19th century wherever Euclid was not used. Indeed, there were several translations into English, including a famous one by Thomas Carlisle, usually credited to Sir David Brewster who oversaw the translation. The book is available only through antiquariat book sellers and in some of the older libraries. It is a must have for those interested in the history of geometry teaching in the United States.

Lewis Carroll, *Euclid and His Modern Rivals,* Dover, New York, 1973.

> Originally published in 1879, with a second edition in 1885, this book argues, in dialogue form, against the replacement of Euclid by numerous other then modern geometry textbooks at the elementary level. Carroll, best known for his Alice books, was a mathematician himself and had taught geometry to schoolboys for almost a quarter of a century when he published the book, which has recently been reprinted by Dover.

David Eugene Smith, *The Teaching of Geometry,* Ginn and Company, Boston, 1911.

> This is not a history book *per se,* but it is of historical interest in a couple of ways. First, it includes a brief history of the subject. Second, it gives a view of the teaching of geometry in the United States at the beginning of the twentieth century. It is currently out of print, but might be available in the better university libraries.

Julian Lowell Coolidge, *A History of Geometrical Methods,* Oxford University Press, 1940.

> This is a rather advanced history of the whole of geometry requiring a knowledge of abstract algebra and the calculus. Publication was taken over by Dover in 1963 and it remains in print.

Felix Klein, *Famous Problems of Elementary Geometry,* Dover.

Wilbur Richard Knorr, *The Ancient Tradition of Geometric Problems,* Dover.

> There are several books on the geometrical construction problems and the proofs of their impossibility. Klein was a leading mathematician of the 19th century, noted for his fine expositions. The book cited is a bit dated, but worth looking into. Knorr is a professional historian of mathematics, whence I would expect more interpretation and analysis and less mathematics from him; I haven't seen his book.

Robert Bonola, *Non-Euclidean Geometry,* Open Court Publishing Company, 1912.

> Republished by Dover in 1955 and still in print in this edition, Bonola is the classic history of non-Euclidean geometry. It includes translations of the original works on the subject by János Bolyai and Nikolai Lobachevsky.

Marvin Jay Greenberg, *Euclidean and Non-Euclidean Geometries; Development and History,* W.H. Freeman and Company, San Francisco, 1974.

> This textbook serves both as an introduction to and a history of non-Euclidean geometry. It contains numerous exercises. The book is currently in its third edition and remains in print.

10 Calculus

Carl Boyer, *History of Analytic Geometry,* The Scholar's Bookshelf, Princeton Junction (NJ), 1988.

—, *The History of the Calculus and Its Conceptual Development,* Dover, New York, 1959.

> These are two reprints, the former from articles originally published in the now defunct journal *Scripta Mathematica* in 1956 and the second published in book form in 1949. Both books discuss rather than do mathematics, so one gets the results but not the proofs of a given period.

Margaret L. Baron, *The Origins of the Infinitesimal Calculus,* Pergamon Press, Oxford, 1969.

> This is a mathematically more detailed volume than Boyer.

C.H. Edwards, Jr., *The Historical Development of the Calculus*, Springer-Verlag, New York, 1979.

> This is a yet more mathematically detailed exposition of the history of the calculus complete with exercises and 150 illustrations.

Judith V. Grabiner, *The Origins of Cauchy's Rigorous Calculus*, MIT Press, Cambridge (Mass), 1981.

> Today's formal definitions of limit, convergence, etc. were written by Cauchy. This book discusses the origins of these definitions. Most college students come out of calculus courses with no understanding of these definitions; they can neither explain them nor reproduce them. Hence, one must consider this a history of advanced mathematics.

11 Women in Science

Given the composition of this class[18], I thought these books deserved special mention. Since women in science were a rare occurrence, there are no unifying scientific threads to lend some structure to their history. The common thread is not scientific but social— their struggles to get their feet in the door and to be recognised. From a masculine point of view this "whining" grows tiresome quickly, but the difficulties are not imaginary. I've spoken to female engineering students who told me of professors who announced women would not get good grades in their classes, and Julia Robinson told me that she accepted the honour of being the first woman president of the American Mathematical Society, despite her disinclination to taking the position, because she felt she owed it to other women in mathematics.

Several books take the struggle to compete in a man's world as their main theme. Some of these follow.

H.J. Mozans, *Women in Science, with an Introductory Chapter on Woman's Long Struggle for Things of the Mind*, MIT Press, Cambridge (Mass), 1974.

> This is a facsimile reprint of a book originally published in 1913. I found some factual errors and thought it a bit enthusiastic.

P.G. Abir-Am and D. Outram, *Uneasy Careers and Intimate Lives; Women in Science, 1789 - 1979*, Rutgers University Press, New Brunswick, 1987.

> Publishing information is for the paperback edition. The book is strong on the struggle, but says little about the science done by the women.

---

[18] Mostly female.

H.M. Pycior, N.G. Stack, and P.G. Abir-Am, *Creative Couples in the Sciences,* Rutgers University Press, New Brunswick, 1996.

> The title pretty much says it all. Pycior has written several nice papers on the history of algebra in the 19th century. I am unfamiliar with the credentials of her co-authors, other than, of course, noticing that Abir-Am was co-author of the preceding book.

G. Kass-Simon and Patricia Farnes, eds., *Women of Science; Righting the Record,* Indiana University Press, Bloomington,1990.

> This is a collection of articles by different authors on women in various branches of science. The article on mathematics was written by Judy Green and Jeanne LaDuke. Both have doctorates in mathematics, and LaDuke also in history of mathematics. With credentials like that, it is a shame their contribution isn't book-length.

There are a few books dedicated to biographies of women of science in general.

Margaret Alic, *Hypatia's Heritage; A History of Women in Science from Antiquity through the Ninetheenth Century,* Beacon Press, 1986.

> Margaret Alic is a molecular biologist who taught courses on the history of women in science, so this narrative ought to be considered fairly authoritative.

Martha J. Bailey, *American Women in Science; A Biographical Dictionary,* ABC-CLIO Inc., Santa Barbara, 1994.

> As the title says, this is a biographical dictionary of women scientists— including some still living, but limited to Americans. The entries are all about one two-column page in size, with bibliographic references to ternary sources. The author is a librarian.

Marilyn Bailey Ogilvie, *Women in Science; Antiquity through the Nineteenth Century,* MIT Press, Cambridge (Mass), 1986.

> Probably the best all-round dictionary of scientific womens' biography.

Sharon Birch McGrayne, *Nobel Women in Science; Their Lives, Struggles and Momentous Discoveries,* Birch Lane Press, New York, 1993.

> There being no Nobel prize in mathematics, this book is only of tangential interest to this course. It features chapter-length biographies of Nobel Prize winning women.

Edna Yost, *Women of Modern Science,* Dodd, Mead and Company, New York, 1959.

> The book includes 11 short biographies of women scientists, none of whom were mathematicians.

Lois Barber Arnold, *Four Lives in Science; Womens' Education in the Nineteenth Century,* Schocken Books, New York, 1984.

This book contains the biographies of 4 relatively obscure women scientists and what they had to go through to acquire their educations and become scientists. Again, none of them were mathematicians.

There are also more specialised collections of biographies of women of mathematics.

Lynn M. Osen, *Women in Mathematics,* MIT Press, Cambridge (Mass), 1974.

Oft reprinted, this work contains chapter-sized biographies of a number of female mathematicians from Hypatia to Emmy Noether.

Miriam Cooney, ed., *Celebrating Women in Mathematics and Science,* National Council of Teachers of Mathematics, Reston (Virginia), 1996.

This book is the result of a year-long seminar on women and science involving classroom teachers. The articles are short biographical sketches written by the teachers for middle school and junior high school students. They vary greatly in quality and do not contain a lot of mathematics. The chapter on Florence Nighingale, for example, barely mentions her statistical work and does not even exhibit one of her pie charts. Each chapter is accompanied by a nice woodcut-like illustration.

Charlene Morrow and Teri Perl, *Notable Women in Mathematics; A Biographical Dictionary,* Greenwood Press, Westport (Conn.), 1998.

This is a collection of biographical essays on 59 women in mathematics from ancient to modern times, the youngest having been born in 1965. The essays were written for the general public and do not go into the mathematics (the papers average 4 to 5 pages in length) but are informative nonetheless. Each essay includes a portrait.

There are quite a few biographies of individual female scientists. Marie Curie is, of course, the most popular subject of such works. In America, Maria Mitchell, the first person to discover a telescopic comet (i.e., one not discernible by the naked eye), is also a popular subject. Florence Nightingale, "the passionate statistician" who believed one could read the will of God through statistics, is the subject of several biographies— that make no mention of her mathematical involvement. Biographies of women of mathematics that unflinchingly acknowledge their mathematical activity include the following.

Maria Dzielska, *Hypatia of Alexandria,* Harvard University Press, 1995.

This is a very scholarly account of what little is known of the life of Hypatia. It doesn't have too much to say about her mathematics, citing but not reproducing a list of titles of her mathematical works.

Nonetheless, the book is valuable for its debunking a number of myths about the subject.

Doris Langley Moore, *Ada, Countess of Lovelace, Byron's Legitimate Daughter,* John Murray, London, 1977.

Dorothy Stein, *Ada; A Life and a Legacy,* MIT Press, Cambridge (Mass), 1985.

Joan Baum, *The Calculating Passion of Ada Byron,* Archon Books, Hamden (Conn), 1986.

Betty Alexandra Toole, *Ada, the Enchantress of Numbers; A Selection from the Letters of Lord Byron's Daughter and Her Description of the First Computer,* Strawberry Press, Mill Valley (Calif), 1992.

I've not seen Moore's book, but do not recommend it[19]. For one thing, I've read that it includes greater coverage of Ada Byron's mother than of Ada herself. For another, Dorothy Stein, in defending the publication of her own biography of Ada Byron so soon after Moore's, says in ther preface, "... a second biography within a decade, of a figure whose achievement turns out not to deserve the recognition accorded it, requires some justification. My study diverges from Mrs. Moore's in a number of ways. The areas she felt unable to explore— the mathematical, the scientific, and the medical— are central to my treatment". A psychologist with a background in physics and computer science, Stein is the only one of Ada's biographers with the obvious credentials to pass an informed judgment on Ada's scientific prowess. And her judgment is very negative.

The romantic myth of a pretty, young girl pioneering computer science by writing the first ever computer program has proven far too strong to be exploded by the iconoclastic Stein. According to the blurb on the dust jacket, "Unlike recent writers on the Countess of Lovelace, Joan Baum does justice both to Ada and to her genuine contribution to the history of science". Of course, an author cannot be blamed for the hype on the dust jacket and Baum is no doubt innocent of the out and out false assertion that "Ada was the first to see from mechanical drawings that the machine, in theory, could be programmed". "The machine" in question is Babbage's analytical engine and was designed expressly for the purpose of being programmed. In any event, Baum is a professor of English and her mathematical background is not described. Approach this book with extreme caution, if at all.

---

[19] The referee, whose comments themselves often display a great deal of respect for authority, admonished me for this remark. However, in the real world, one must decide whether or not to expend the effort necessary to consult one more reference. In the present case, Stein's credentials are impeccable, her writing convincing, and her comments say to me that Moore's book contains nothing of interest to me. This suffices for me.

Toole's book consists of correspondence of Ada Byron "narrated and edited" by a woman with a doctorate in education. The editing is fine, but the narration suspect. At one point she describes as sound a young Ada's speculation on flying— by making herself a pair of wings! I for one have seen enough film clips of men falling flat on their faces after strapping on wings to question this evaluation of Ada's childhood daydreams. Approach with caution.

Louis L. Bucciarelli and Nancy Dworsky, *Sophie Germain; An Essay in the History of the Theory of Elasticity,* D. Reidel Publishing Company, Dordrecht, 1980.

This is an excellent account of the strengths and weaknesses of a talented mathematician who lacked the formal education of her contemporaries.

Sofya Kovalevskaya, *A Russian Childhood,* Springer-Verlag, New York, 1978.

Pelageya Kochina, *Love and Mathematics: Sofya Kovalevskaya,* Mir Publishers, Moscow, 1985. (Russian original: 1981.)

Ann Hibler Koblitz, *A Convergence of Lives; Sofia Kovalevskaia: Scientist, Writer, Revolutionary,* Birkhäuser, Boston, 1983.

Roger Cooke, *The Mathematics of Sonya Kovalevskaya,* Springer-Verlag, New York, 1984.

Before Emmy Noether, Sofia Kovalevskaya was the greatest woman mathematician who had ever lived. She was famous in her day in a way unusual for scientists. *A Russian Childhood* is a modern translation by Beatrice Stillman of an autobiographical account of her youth first published in 1889 in Swedish in the guise of a novel and in the same year in Russian. Over the next several years it was translated into French, German, Dutch, Danish, Polish, Czech, and Japanese. Two translations into English appeared in 1895, both published in New York, one by The Century Company and one by Macmillan and Company. Each of these volumes also included its own translation of Charlotte Mittag-Leffler's biography of her. The original translations are described by the new translator as being "riddled with errors", which explains the need for the new edition, which also includes a short autobiographical sketch completing Kovalevskaya's life story and a short account of her work by Kochina, to whom, incidentally, the book is dedicated.

The volumes by Kochina and Koblitz are scholarly works. Kochina was head of the section of mathematical methods at the Institute of Problems of Mechanics of the Soviet Academy of Sciences and is also known for her work in the history of mathematics. Koblitz's areas of expertise are the history of science, Russian intellectual history, and women in science. Both women are peculiarly qualified to write a biography of Kovalevskaya. Kochina's book actually includes some mathematics.

Cooke's book discusses Kovalevskaya's mathematical work in detail, placing it in historical context. He includes biographical information as well. This book is quite technical and not for the weak at heart.

Auguste Dick, *Emmy Noether, 1882 - 1935,* Birkhäuser, Boston, 1981.

This is a short biography of the greatest woman mathematician to date. It includes three obituaries by such mathematical notables as B.L. van der Wærden, Hermann Weyl, and P.S. Alexandrov.

Tony Morrison, *The Mystery of the Nasca Lines,* Nonesuch Expeditions Ltd., Woodbridge (Suffolk), 1987.

The author is an English man and not the African American poetess (Ton*i*). The Nasca Lines are lines laid out by prehistoric Indians on a high, dry plateau in Peru. The book has much information on these lines and Maria Reiche's studies of them, as well as biographical information on Reiche. Reiche studied mathematics in Germany before moving to Peru and making a study of the lines her life's work. The book has lots of photographs.

The Nasca Lines and Maria Reiche have been the subjects of televised science specials. According to these, her specific astronomical interpretations of the lines are in dispute, but her demonstrations of the utterly simple geometric constructions that can be used to draw the figures accompanying the lines obviate the need to assume them the work of ancient astronauts *à la* Erich von Däniken. I don't recall this being in the book, which I found at a local library.

Constance Reid, *Julia; A Life in Mathematics,* Mathematical Association of America, 1996.

This is a very pleasant little volume on the life and work of Julia Robinson. It contains an "autobiography" actually written by Robinson's sister Constance Reid, as well as three articles on her mathematical work written by her friend Lisl Gaal and her friends and collaborators Martin Davis and Yuri Matijasevich.

Constance Reid has written a number of popular biographies of mathematicians. She is not a mathematician herself, but had access to mathematicians, in particular, Julia Robinson and her husband Raphæl.

12 Miscellaneous Topics

F.N. David, *Games, Gods and Gambling; A History of Probability and Statistical Ideas,* Dover.

Jacob Klein, *Greek Mathematical Thought and the Origin of Algebra,* Dover.

I haven't seen these books which are listed on Dover's website.

Petr Beckmann, *A History of* π, Golem Press, 1971.

Lennart Berggren, Jonathan Borwein, and Peter Borwein, *Pi: A Source Book*, Springer-Verlag, New York, 1997.

> This is a delightful book on π, covering everything from early esti-
> mates to the attempt by the Indiana legislature to pass a law making
> the number rational. The book has been republished by St. Martin's
> Press and is currently in print in a paperback edition by St. Martin's.
> As the title says, the book by Berggren *et al.* is a source book, con-
> sisting of a broad selection of papers on π of varying levels of diffi-
> culty. The book is not annotated, several papers in Latin, German
> and French are untranslated, some of the small print is illegible (too
> muddy in Lindemann's paper on the transcendence of π and too faint
> in Weierstrass's simplification), and the reader is left to his or her own
> devices. Nonetheless, there is plenty of material accessible to most stu-
> dents. The book is currently in its third (2004) edition. There are other
> books on π, as well as books on e, the golden ratio, and i, but these
> books are particularly worthy of one's attention.

Elisha S. Loomis, *The Pythagorean Proposition. Its Demonstrations Analyzed and Classified and Bibliography of Sources for Data of the Four Kinds of "Proofs"*, 2nd. ed., Edwards Brothers, Ann Arbor, 1940.

> Originally published in 1927, the book received the endorsement of
> the National Council of Teachers of Mathematics when it published
> a reprint in 1972. I've not seen the book, and paraphrase my friend
> Eckart Menzler-Trott: 370 proofs are analysed in terms of being alge-
> braic (109), geometric (255), quaternionic (4), or dynamic (2). He jok-
> ingly states that it is the Holy Book of esoteric Pythagoreans, having
> got the book through a religious web site. Indeed, I myself purchased
> a couple of biographies of Pythagoras at a religious bookstore, and
> not at a scientific bookseller's.

13 Special Mention

Anon, ed., *Historical Topics for the Mathematics Classroom*, National Council of Teachers of Mathematics, Washington, D.C., 1969.

> This is a collection of chapters on various topics (numbers, compu-
> tation, geometry, etc. up to and including calculus) with historical
> information on various aspects of these topics. The discussions do not
> include a lot of mathematical detail (e.g., it gives the definition and
> graph of the quadratrix, but does not derive the equations and show
> how to square the circle with it). Nonetheless, if one is interested in

using history in the classroom, it is a good place to start looking for ideas on just how to do so.

Ludwig Darmstædter, *Handbuch zur Geschichte der Naturwissenschaften und der Technik,* 2nd enlarged edition, Springer-Verlag, Berlin, 1908.

Claire L. Parkinson, *Breakthroughs; A Chronology of Great Achievements in Science and Mathematics 1200 - 1930,* GK Hall and Company, Boston, 1985.

Darmstædter's book is a carefully researched 1070 page chronology of all of science up to 1908. It exists in an authorised reprint by Kraus Reprint Co., Millwood, NY, 1978, and thus ought to be in any respectable American university library.

Parkinson's book is a modern replacement for Darmstædter's. It brings one a bit more up to date, but starts a lot later. Parkinson compiled her dates from more secondary sources than did Darmstædter, and her book is probably best viewed more as a popularisation than as a scholarly reference work. On the other hand, her book does have an extensive bibliography, which Darmstædter's does not. Moreover, neither book is illustrated and they ought not to be confused with some more recent coffee table publications on the subject.

Chronologies are not all that useful, a fact possibly first made manifest by the failure of Darmstædter's massive effort to have had an effect on the history of science[20]. One limitation of the usefulness of such a volume is the breadth of coverage for a fixed number of pages: more exhaustive coverage means shorter entries. For example, we read in Darmstædter that in 1872 Georg Cantor founded "the mathematical theory of manifolds (theory of point sets)", i.e. Cantor founded set theory in 1872. What does this mean? Is this when Cantor started his studies of set theory, when he published his first paper on the subject, a date by which he had most of the elements of the theory in place, or...? Similarly, we read that in 250 A.D., "Diophantus of Alexandria freed arithmetic from the bonds of geometry and founded a new arithmetic and algebra on the Egyptian model". What does this mean? How does the algebra founded by Diophantus compare with the geometric algebra of Euclid, the later algebra of al-Khwarezmi, or the "letter calculus" [21] by Viète in 1580? To answer these questions, one must go elsewhere.

Another problem concerns events we do not have the exact dates of. In compiling a chronology, does one include only those events one can date exactly, or does one give best guesses for the uncertain ones? Darmstædter has done the latter, as evidenced by his dating of Diophantus at 250 AD. Unfortunately, he did not write "c. 250" to indicate this to be only an approximation. Is the year 1872 cited for Cantor an exact date or an estimate, perhaps a

---

[20] Helge Krogh, *An Introduction to the Historiography of Science,* Cambridge University Press, Cambridge, 1987, pp. 17 and 175.

[21] I.e., the use of letters as variables.

midpoint in Cantor's career? Once again one has to look elsewhere for the answer.

And, of course, a problem with Darmstædter or any older reference is that later research may allow us to place questionably dated events more exactly in time. It may uncover events that were unknown and thus left out of the chronology. Darmstædter himself cites the 1906 discovery of *The Method* of Archimedes by J.L. Heiberg, a result that Darmstædter could not have included in his first edition. And such research may correct other errors: Darmstædter cites Euclid of Megara as the author of the *Elements*, a common misidentification we discussed in Chapter 1.

Ivor Grattan-Guinness, ed., *Companion Encyclopedia of the History and Philosophy of the Mathematical Sciences,* 2 vols., Routledge, London and New York, 1992.

> With over 1700 pages not counting the end matter, these two volumes give a very broad but shallow coverage of the whole of mathematics. It has a useful annotated bibliography as well as a chronology.

Augustus de Morgan, *A Budget of Paradoxes,* 2 vols., 2nd ed., Open Court Publishing Company, Chicago and London, 1915.

> The first edition was published in 1872, edited by de Morgan's widow Sophia. The second edition was edited by David Eugene Smith. Some later printings of the second edition appeared unter the title *An Encyclopædia of Eccentrics.*
> The book is an amazing bit of odds and ends— anecdotes, opinion pieces, and even short reviews, some dealing with mathematical subjects and some not. Smith's description of the work as a "curious medley" and reference to its "delicious satire" sum it up nicely.

14 Philately

In connexion with the final chapter of this book, I cite a few references on mathematics and science on stamps.

W.J. Bishop and N.M. Matheson, *Medicine and Science in Postage Stamps,* Harvey and Blythe Ltd., London, 1948.

> This slim volume written by a librarian of a medical museum and a surgeon contains a short 16 page essay, 32 pages of plates sporting 3 to 6 stamps each, a 3 page bibliography, and 23 pages of mini-biographies of the physicians and scientists depicted on the stamps, together with years of issue and face values of the stamps.

R.W. Truman, *Science Stamps*, American Topical Association, Milwaukee (Wisconsin), 1975.

> This is a rambling account of science on stamps with 7 pages densely covered with images of stamps. The text is divided into three parts, the first being split into shorter chapters on Physicists, Chemistry, Natural History, Medicine, and Inventors; the second with special chapters on The Curies, Louis Pasteur, Alexander von Humboldt, Albert Schweitzer, and Leonardo da Vinci; and the third with chapters on the stamps of Poland, France, Germany, Russia, and Italy. There are also checklists of stamps by name, country, and scientific discipline. These checklists give years of issue and catalogue numbers from the American *Scott* postage stamp catalogue.

William L. Schaaf, *Mathematics and Science; An Adventure in Postage Stamps*, National Council of Teachers of Mathematics, Reston (Virginia), 1978.

> This is the earliest book in my collection to deal primarily with mathematics. It has a narrative history of mathematics illustrated by postage stamps, some reproduced in colour. This is supplemented by two checklists, one by scientist and one by subject. The checklists are based on the American catalogue.

Peter Schreiber, *Die Mathematik und ihre Geschichte im Spiegel der Philatelie*, B.G. Teubner, Leipzig, 1980.

> This slim East German paperback contains the customary short history of mathematics illustrated by 16 pages of not especially well printed colour plates featuring about a dozen stamps each. Its checklist is by country, but there is a name index that allows one to look up an individual. The checklist is based on the East German *Lipsius* catalogue, which is quite rare. When I visited the library of the American Philatelic Association some years ago, they didn't have a complete catalogue.

Robert L. Weber, *Physics on Stamps*, A. S. Barnes and Company, Inc., San Diego, 1980.

> This book concerns physics, not mathematics, but some of the stamps are of mathematical interest. The book makes no attempt at complete coverage. There is no checklist and the narrative is a sequence of topical essays illustrated by postage stamps, all in black and white despite the promise of colour on the blurb on the dust jacket.

Hans Wussing and Horst Remane, *Wissenschaftsgeschichte en Miniature*, VEB Deutscher Verlag der Wissenschaften, Berlin, 1989

> This very attractive volume consists of essays on the development of science through the ages, each page illustrated with 4 to 6 stamps. Wussing is both an historian and a philatelist.

Robin J. Wilson, *Stamping Through Mathematics*, Springer-Verlag, New York, 2001.

This slim volume consists of a collection of short 1 page essays on various topics in the history of mathematics, each illustrated by a page of 6 to 8 beautiful oversize colour reproductions of appropriate stamps.

3

Foundations of Geometry

1 The Theorem of Pythagoras

It has been known for some time that the Pythagorean Theorem did not originate with Pythagoras. He is credited, however, with having given the first proof thereof— a credit not in serious doubt as far as Western mathematics is concerned[1]. There are other early proofs from China and India, with the Chinese claiming priority by a period of time that is not determined too exactly by the references at hand.[2] But, except for ethnic bragging rights— and I am neither Greek nor Chinese nor Hindi— the earliest authorship is unimportant. What I wish to discuss is not the original discoverer of the proof, but the original proof and form of the result.

I have on the shelves of my library a curious volume by a Dr. H.A. Naber, a secondary school teacher in Hoorn (Netherlands), published in 1908 under

---

[1] To be sure, there is room for doubt. That Pythagoras was the European author of the theorem is a tradition, not a documented fact. However, the tradition is strong and, there being no arguable alternative hypothesis, one simply accepts it.

[2] For example, Mikami's *Mathematics in China and Japan* places Pythagoras "six long centuries" after the proof given in the *Chou-pei Suan-ching*, while Joseph's *The Crest of the Peacock* says, "While it is no longer believed that this treatise predates Pythagoras by five centuries, it is still thought likely that it was composed before the time of the Greek mathematician". Lǐ and Dù's *Chinese Mathematics; A Concise History* does not commit to a pre-Pythagorean date for the treatment of the Pythagorean Theorem in the *Nine Chapters*, while Joseph Dauben, in a Festschrift for Hans Wussing (full citation in footnote 8 in the Bibliography) dates the composition to no later than 1100 B.C. and possibly as early as the 27th century. The problem is that in 213 B.C., the emperor decreed all books burned and all scholars buried. The emperor did not survive long and searches were made and unfound classics rewritten and one cannot be certain of which passages might have included newer material.

the lengthy and, to modern tastes, pretentious title,

Das
Theorem des Pythagoras
wiederhergestellt in seiner
ursprünglichen Form und betrachtet als
Grundlage der ganzen
Pythagoreanischen Philosophie.

This translates to *The Theorem of Pythagoras, Restored to Its Original Form and Considered as the Foundation of the Entire Pythagorean Philosophy.* The book is a mixture of mathematics and unbridled speculation. But it is interesting, and it does raise an interesting question: what is the original form of the Pythagorean Theorem?

Naber cites the existence of 70 proofs of the Pythagorean Theorem, and considerably more are catalogued in the book by Loomis cited in the bibliograpy. But for many centuries schoolboys in the West had to struggle with Euclid's proof and, unless they became mathematicians, would never see another. While Edna St. Vincent Millay may rhapsodise on how "Euclid alone has looked on Beauty bare", keener critics have had less kind things to say. Naber offers a few choice quotations, including Arthur Schopenhauer in *The World as Will and Representation*: Des Eukleides stelzbeiniger, ja hinterlistiger Beweis verlässt uns beim Warum. (Very roughly: Euclid's stiltwalking, indeed cunning proof leaves us wondering why.) I'm not sure if Schopenhauer's "why" is asking why the result is true, or why Euclid gave the proof that he did. For, Euclid's proof is surely one of the most complex and difficult ones on record, and it can't be said to yield much insight.

Let us briefly review this proof, which is essentially what Book I of the *Elements* builds up to. Euclid begins Book I with some basics— some congruence properties of triangles and basic constructions (e.g., angle bisections)— and then gets into the theory of parallel lines eventually proving Proposition 36: *Parallelograms which are on equal bases and in the same parallel lines are equal to one another.* In other words, for parallelograms, equal base and equal height mean equal area. He then repeats the exercise to show the same to hold for triangles. After showing that one can construct a square on a given side, he is now ready to prove the Pythagorean Theorem.

Sketch of Euclid's Proof. Let ABC be a right triangle, with the right angle at C and draw the figure so the triangle is resting on the hypotenuse. Put squares on each of the sides and drop the perpendicular from C to the far side of the opposite square as in *Figure 1*. Draw the lines connecting C to the vertex D and B to the vertex I.

The area of triangle CAD is the same as that of ADK since they share the base AD and are both trapped between the parallels AD and CJ. Thus CAD is half the area of $ADJK$. Similarly, IAB is half the area of $ACHI$.

But it turns out that IAB and CAD are congruent: IA and CA are sides of a square, whence equal. The same holds of AB and AD. Finally, the angles between these corresponding sides are each the sum of a right angle and $\angle CAB$; thus they are also equal. Now we are just about finished because $ADJK$ has area double that of CAD, which is double that of IAB, which is the area of $ACHI$. In a similar manner one sees $KJEB$ to equal $CBFG$, whence the area of $ADEB$ is the sum of $CBFG$ and $ACHI$.   □

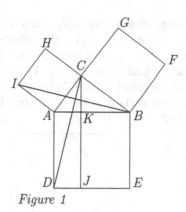

Figure 1

Euclid's proof is nice in that it does show how the areas of the small squares fit into the large square— but less complicated proofs do the same, albeit without such simple regions. Joseph Dauben[3] presents the Chinese proof reproduced in *Figure 2*, below. This is a simple dissection proof whereby one cuts the large inscribed square up and reassembles it to make the two smaller squares. Subtractive proofs also exist. Naber cites the Dutch writer Multatuli (real name: Douwes Dekker) for what the latter declared to be the simplest possible proof and which I reproduce as *Figure 3*, below.

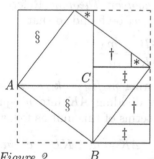

Figure 2   B

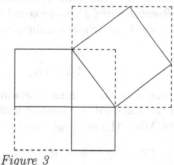

Figure 3

Naber also notes that this proof was known to the Hindu mathematicians. The most famous Indian proof, however, has got to be that of Bhaskara, which I reproduce in full in *Figure 4*.

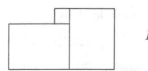

Behold!

Figure 4

---

[3] Cf. footnote 8 of the Bibliography. Cf. also note 2, above.

Nowadays we add the explanation that, if a, b are the legs of the right triangle, with b the shorter, and c the hypotenuse, the first half of the figure is $c^2 = 4\left(\frac{1}{2}ab\right) + (a-b)^2$ and the right half is $2ab + (a-b)^2$, whence equal to c^2. But it is also equal to $a^2 + b^2$, as can be seen by extending the leftmost vertical side of the small square downward.

A final variant is the geometric form of comparing $(a+b)^2$ with c^2 as in *Figure 5*.

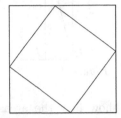 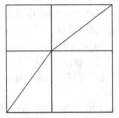

Figure 5

With all these simple proofs, the question to ask is why Euclid gave the proof that he did. Thomas Heath, in his definitively annotated edition of the *Elements*, suggests that Euclid may have given a correct version of an earlier Pythagorean proof that was invalidated by the discovery of irrational numbers. Arthur Gittleman's textbook[4] on the history of mathematics elaborates on this and, since the point is not made in most other textbooks, I repeat it here.

Reconstructed Pythagorean Proof of the Pythagorean Theorem. Referring to *Figure 1*, let whole numbers $m_1, m_2, m_3, n_1, n_2, n_3$ be found so that

$$\frac{AC}{AK} = \frac{m_1}{n_1}, \quad \frac{BC}{AK} = \frac{m_2}{n_2}, \quad \frac{AB}{AK} = \frac{m_3}{n_3}.$$

Letting n be the common denominator, and writing $m_i/n_i = k_i/n$, we see that we can find a common unit of measure by dividing AK into n equal parts. Write the integral lengths of these lines in terms of this unit as follows:

$$a = BC = k_2, \quad b = AC = k_1, \quad c = AB = k_3, \quad AK = n, \quad KB = c - n.$$

The line CK subdivides the triangle ABC into triangles similar to itself: ABC is similar to ACK and CBK. To see the first, note that

$$\angle CAB = \angle KAC$$

since they physically coincide. Also,

$$\angle ACB = \angle AKC,$$

since they are right angles. The remaining angles must also be equal and similarity is established. But then

---

[4] Arthur Gittleman, *History of Mathematics*, Merrill Publishing Company, 1975.

$$\frac{AC}{AK} = \frac{AB}{AC}, \quad \text{i.e.,} \quad \frac{b}{n} = \frac{c}{b}.$$

These are ratios of whole numbers, whence they can be multiplied to yield $b^2 = nc$, i.e. the area of the square $ACHI$ equals that of the rectangle $ADJK$ as before.

Similarly, $CBFG$ equals $KJEB$ and $a^2 + b^2 = nc + (c-n)c = c^2$.   □

This proof is a bit more memorable than Euclid's and today, with our willingness to multiply and divide arbitrary real numbers, the proof is valid as soon as we remove the step yielding integers. We just set

$$a = BC, \quad b = AC, \quad c = AB, \quad d = AK, \quad c - d = KB,$$

cite the similarity of the triangles, and conclude

$$\frac{b}{d} = \frac{c}{b}, \quad \frac{a}{c-d} = \frac{c}{a}$$

whence $b^2 = cd, a^2 = c(c-d)$, and the areas of the two rectangular pieces of the large square are those of the corresponding small squares.

Depending as it does on the concept of similarity, to present the proof in this manner, Euclid would first have to develop the theory of proportions and then the theory of similar triangles. The theory of proportions, however, is much more abstract than anything in the first book of the *Elements* and, like the use of the axiom of choice today, there is a premium on its avoidance.

Whatever the reason or configuration of reasons, Euclid chose to give the intricate but elementary proof he gave and that is that.

Euclid followed his proof of the Pythagorean Theorem with a proof of its converse: *If in a triangle the square on one of the sides is equal to the squares on the remaining two sides of the triangle, the angle contained by the remaining two sides of the triangle is right.* The proof is a fine example of minimalism. It establishes what is needed and no more. One starts with ABC, supposing the square on AB being the sum of the squares on AC and BC, and draws a line from C perpendicular to BC to a point D equal in distance to AC as in *Figure 6*. Because $AC = DC$ and $CB = CB$, two sides of the triangles ABC and DBC are equal. Now the square on BD is the sum of the squares on BC and CD by the Pythagorean Theorem and the square on AB is the sum of the squares on CA and BC by assumption, whence the square on BD equals the square on AB. But then BD equals AB and the triangles ABC and DBC are congruent, whence $\angle ACB$ is a right angle.

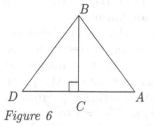

Figure 6

Later, in Propositions 12 and 13 of Book II, he generalises both the Pythagorean Theorem and its converse to an arbitrary triangle by providing error terms in what we now recognise as geometric forms of the Law of Cosines.

In Book VI, Euclid offers another generalisation of the Pythagorean Theorem in Proposition 31: *In right-angled triangles the figure on the side subtending the right angle is equal to the similar and similarly described figures on the sides containing the right angle.* The proof of this is not easy to follow because Euclid offers a curious circumlocution involving things being in "duplicate ratio". Going into this proof he has Proposition VI-20 at his disposal, which we may restate as follows: *Let P_1, P_2 be similar polygonal figures with sides s_1, s_2 corresponding under the similarity. Then P_1 is to P_2 as the square on side s_1 is to the square on side s_2.* In other words, the ratio of the area of P_1 to P_2 is the same as the ratio of the squares of the lengths of s_1 and s_2.

Modulo a change of notation, Euclid draws the figure on the right, where CD is perpendicular to AB and α, β, γ denote the areas of the similar polygons. Now, by VI-20,

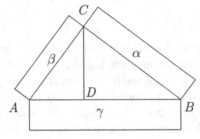

Figure 7

$$\frac{\beta}{\gamma} = \frac{AC^2}{AB^2}, \quad \frac{\alpha}{\gamma} = \frac{BC^2}{AB^2},$$

whence

$$\frac{\alpha + \beta}{\gamma} = \frac{BC^2 + AC^2}{AB^2}. \tag{1}$$

By the same token,

$$\frac{ACD}{ABC} = \frac{AC^2}{AB^2}, \quad \frac{CBD}{ABC} = \frac{BC^2}{AB^2}$$

and

$$\frac{ACD + CBD}{ABC} = \frac{\alpha + \beta}{\gamma}.$$

But $ACD + CBD = ABC$, whence $1 = \frac{\alpha+\beta}{\gamma}$ and $\gamma = \alpha + \beta$. But this is VI-31 and we are finished.

The Pythagorean Theorem is obviously a special case of this theorem and, assuming the results it depended on did not depend on the Pythagorean Theorem itself, we have another proof of the latter. Alternatively, we can avoid the last step in the proof by appealing to the Pythagorean Theorem right after proving (1). From the assumption $AB^2 = AC^2 + BC^2$, one goes from (1) to $\frac{\alpha+\beta}{\gamma} = 1$ to $\gamma = \alpha + \beta$.

And this brings us back to Naber and his hunt for the theorem in its original form. According to him, we know the Theorem is valid for any similar polygons constructed on the sides of the right triangle. We also know that squares will yield no insight (else why would there be over 70 proofs?). Therefore let us place similar triangles on the sides instead of squares. In fact, let them be similar to ABC itself as in *Figure 8*, below. But now he recalls Euclid's Proposition VI-8 asserting the triangles we've been considering all along— the ones obtained by dropping the perpendicular from C to AB— are similar to ABC[5]. Thus he sees as the *ur*form of the Pythagorean Theorem the equation of *Figure 9*.

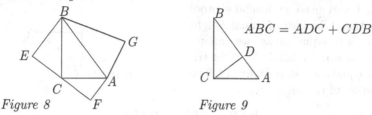

Figure 8   Figure 9

It may or may not take a moment's thought to see that the desired equation $AGB = CEB + AFC$ of *Figure 8* is the same as the equation of *Figure 9*, but it is: triangle AGB is just the reflexion across AB of triangle ACB of *Figure 9*, CEB the reflexion across BC of CDB, and AFC the reflexion across AC of ADC.

So, is Naber right? Is *Figure 9* the *ur*form of the Pythagorean Theorem, and, if so, what about Figure 10, below? In fact, can't we generalise this further by dropping the requirement that $\angle ACB$ be a right angle? Why the right angles at C and D in *Figure 9*? The reason is that there are two components to the proof of the Pythagorean Theorem as represented in *Figure 9* and only one is present in *Figure 10*. First, of course, is the instance, $ABC = ADC + CDB$, of the axiom asserting the whole is equal to the sum of its parts. Second, however, is the requirement that the polygons placed on the sides of the triangles be similar— and this clearly is not the case for the triangles that would be obtained from *Figure 10* by reflecting ACD across AC, BCD across BC, and ACB across AB. That it is the case for the triangles of *Figure 8* resulting from *Figure 9* follows from Euclid's Proposition VI-8.

Figure 10

But consider this: however immediate to the senses *Figure 9* may be and however quickly we conclude $AGB = AFC + CEB$ in *Figure 8*, this is still not the Pythagorean Theorem. *Figure 9* is only a viable candidate for recognition as the original[6] form if the reduction of the Pythagorean Theorem to *Figure*

---

[5] We actually proved this in the course of reconstructing the Pythagorean proof, above.

[6] Speaking strictly historically, this is bunk: it often happens that the first formulation and proof of a result are over-complicated. But I don't think Naber was seriously interested in the historical accident, but was, rather, interested in find-

9 is relatively simple. The proof given by Euclid can be replaced by a simpler one as follows.

Proof of the Pythagorean Theorem by appeal to Figure 9. Again using Euclid's VI-8, we can conclude from *Figure 9* that the exterior triangles of *Figure 8* are similar and satisfy $AGB = AFC + CEB$. Using the fact (proven in Book I and used in Euclid's proof of I-47) that triangles of equal height and base have equal areas, we can replace the external triangle by right triangles of equal area as in *Figure 11*. The preservation of the area means

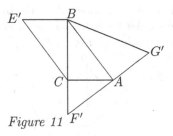

Figure 11

$$AG'B = BE'C + CF'A. \tag{2}$$

The triangles are still similar and we have

$$\frac{AB}{AG'} = \frac{BC}{BE'} = \frac{CA}{CF'}.$$

Call this common ratio λ.

Each triangle is half of the rectangle of which the triangle's hypotenuse is the diagonal. Replacing the triangles by these rectangles we get something like *Figure 7*, rotated. The common ratio λ is the ratio of the lengths of these rectangles to their widths, and is the ratio of the areas of the squares on the lengths to the areas of the rectangles themselves:

$$\frac{AB^2}{AB \cdot AG'} = \frac{BC^2}{BC \cdot BE'} = \frac{CA^2}{CA \cdot CF'} = \lambda.$$

Thus,

$$
\begin{aligned}
AB^2 &= \lambda \cdot AB \cdot AG' \\
&= \lambda [BC \cdot BE' + CA \cdot CF'], \text{(essentially) by (2)} \\
&= \lambda \cdot BC \cdot BE' + \lambda \cdot CA \cdot CF' \\
&= BC^2 + CA^2,
\end{aligned}
$$

as was to be shown.   □

This is certainly simpler and more memorable than Euclid's proof of I-47. But it does depend heavily on similarity, which, by Euclids's proof, the Theorem does not.

---

ing the proof that lies in Plato's world of forms, the one which the 70 proofs he cites are mere human approximations to. If we accept the divinity of Pythagoras, who was the Hyperborean Apollo and had a golden leg to prove it, then it is plausible that the first proof that occurred to *Him* was this platonic ideal of a proof. But such speculation is really taking us off the deep end, so to speak.

It is easy to dismiss Naber out of hand: his great insight is pretty much just Euclid's proof of VI-31, he inflates its importance tremendously, and he surrounds it with everything but the kitchen sink— Pythagorean philosophy, the golden ratio, the great pyramid, etc., etc. Against these most inauspicious trappings there is, however, the fact that this great insight of his is genuinely insightful. No less a light than Georg Pólya[7] also saw fit to comment on the proof of VI-31. His remarks on it, however, were a bit more measured: instead of reading philosophical significance into the proof, he saw in it a useful example for mathematical pædagogy. He viewed proving the Pythagorean Theorem as a problem to be solved. The first step in the solution is to realise that the square shape of the figures erected on the sides is irrelevant. What matters is the similarity of the figures. Thus, one generalises the problem. The next step is the realisation that one can prove the general result if one can prove it for any shape, or similarity class if you will. Finally, comes the realisation that, à la Figure 9, we already know the result for one such set of similar triangles.

2 The Discovery of Irrational Numbers

The most amazing event in the history of Greek mathematics has to have been the discovery of irrational numbers. This was not merely a fact about real numbers, which didn't exist yet. It was a blow to Pythagorean philosophy, one of the main tenets of which was that all was number and all relations were thus ratios. And it was a genuine foundational crisis: the discovery of irrational numbers invalidated mathematical proofs. More than that, it left open the question of what one even meant by proportion and similarity.

The discovery of irrationals is wrapped in mystery. We don't know who discovered them. We don't know exactly when they were discovered. And we don't even know which was the first number recognised to be irrational. Other than the never disputed assertion that the Pythagoreans, who had based so much of their mathematics on the assumption of rationality, discovered the existence of irrational numbers and the belief that they tried to keep the discovery to themselves for a while, all else is legend. And quite a legend it is.

The story goes that Pythagoras was walking down the street listening to the melodious tones coming from the local blacksmith shop. After consulting with the smith, he went home and experimented on strings of various lengths and discovered the most harmonious sounds arose when he plucked strings whose lengths stood in simple ratios to one another. Thus were the seven-stringed lyre and the musical scale invented. Thus too, apparently, was

---

[7] George Polya, "Generalization, specialization, analogy", *American Mathematical Monthly* 55 (1948), pp. 241 - 243, and again in *Induction and Analogy in Mathematics*, Princeton University Press, Princeton, 1954. Pólya was a competent researcher whose interests turned to pædagogy. He is not one to be dismissed lightly.

the foundation of Pythagorean mathematics laid. In early Pythagorean philosophy, all things were numbers and all relations between things were thus numerical ratios, like the relations between harmonious notes on the scale.

Aristotle mentioned that the late Pythagorean Eurytus would determine the number of an object (e.g., a man) by making a picture of the object with pebbles and counting the pebbles. Theophrastus confirms this and Alexander of Aphrodisias expands on it: Eurytus would say, "Suppose the number of man is 250", and illustrate this by smearing plaster on the wall and using 250 pebbles to outline the figure of a man[8].

Specific numbers had their own special properties: 1 generated all other numbers and hence was the number of reason; 2 was the number of opinion— hence the feminine; 3, being composed of unity (1) and diversity (2), was the number of harmony— incidentally (or, therefore) it represented the male; 4 represented the squaring of accounts and so was the number of justice; and $5 = 2 + 3$ was the number of marriage. When the Pythagoreans said *all* was number, they evidently meant it.

And again, if all is number, all relations are ratios of numbers. What one is to conclude from the observation that the relation between man and woman is 3 to 2 is something I would not care to hazard a guess on. In mathematics, however, one can see where this would lead. The Pythagoreans gave primacy to arithmetic, in which field they developed a full theory of proportions, proving things like

$$\frac{a}{b} = \frac{c}{d} \quad \text{iff} \quad \frac{a}{c} = \frac{b}{d}.$$

And in geometry they applied proportions to questions of similarity. The most fundamental geometric consequence of the Pythagorean belief that all was number would be that any two line segments stood in rational relation to one another and hence could be measured by a common unit, i.e. they were *commensurable*. If, for example, segment AB stood in relation 7 : 3 to segment CD, then a segment equalling 1/7th of AB was the same as a third of CD, whence it would serve as a common unit to measure AB and CD by.

The Pythagoreans were not a school of mathematicians in the modern sense. They were a cult and initially were very secretive. We don't know any of their proofs and can only guess. The natural guess is that, as in the reconstructed proof of the Pythagorean Theorem given in the last section, they freely assumed the existence of a common measure in their geometric proofs. As we saw, there are other proofs of the Pythagorean Theorem that can be given that do not depend on this commensurability assumption. A point that is brought out nicely in Arthur Gittleman's textbook[9], but ignored by others, is that this is not always the case. The example he cites is the following (Euclid VI-1).

---

[8] Edward Maziarz and Thomas Greenwood, *Greek Mathematical Philosophy*, Frederick Ungar Publishing Co., New York, 1968, p. 16
[9] Cf. footnote 4.

2.1 Theorem. *Two triangles of equal height are to each other as their bases.*

Numerically stated: the ratio of the areas of two triangles of equal height is the ratio of the lengths of their bases[10].

Unlike the Pythagorean Theorem, this Theorem cannot be given a different proof without some new assumptions[11].

Let us consider the proof of Theorem 2.1 assuming the axiom of commensurability. We may assume the special case (Euclid I-37) which is proven without appeal to commensurability or proportions:

2.2 Lemma. *Two triangles of equal height and equal base are equal.*

That is: triangles of equal heights and equal bases have equal areas.

Proof of Theorem 2.1. Imagine the two triangles as drawn in *Figure 12.*

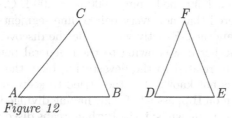

Figure 12

Assume $AB/DE = m/n$ for some whole numbers m, n. Divide AB into m equal subintervals and DE into n such. Since $AB/m = DE/n$, the resulting tiny triangles (Cf. *Figure 13.*) have equal bases.

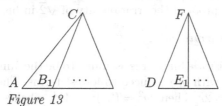

Figure 13

Hence they are of equal areas. Thus

---

[10] The Euclidean tradition is not to use mensuration formulæ like

$$Area = \frac{1}{2} \times Base \times Height,$$

because the Greeks were dealing with *magnitudes* instead of numbers. One historian told me the best way to view magnitudes is as vectors over the rationals— one can add them and multiply them by positive rational numbers, but one cannot multiply or divide them by other magnitudes.

[11] This is not like the independence of the parallel postulate, where one gives a model of geometry in which the postulate is false. In the present case, until one explains what the ratio of two incommensurable line segments is, the statement of the theorem is rendered *meaningless* once such segments are shown to exist.

$$Area(ABC) = m \times (Area(AB_1C)) = m \times (Area(DE_1F))$$

$$= m \times \left(\frac{1}{n} \times Area(DEF) \right),$$

and $ABC/DEF = m/n = AB/DE$. $\qquad\qquad\qquad\qquad\qquad\qquad\qquad$ □

The discovery of irrational numbers doesn't completely invalidate the proof given; it merely narrows its range of applicability: the proof works for pairs of triangles of equal height whose bases stand in some rational proportion. For other triangles, however, the statement is not only unproven but meaningless[12]. Only after Eudoxus explained what is meant by the ratio of incommensurable lines would the full result become an open problem amenable to solution. But this discussion belongs to the next section; for now we have the discovery of irrationals to discuss.

The Pythagoreans flourished from $c.$ 600 to $c.$ 400 B.C. Some time in that period they discovered that not every pair of line segments stood in rational proportion to one another. Exactly who made the discovery and how he did it is unknown. Whether it was owing to their general penchant for secrecy, or was more directly related to the devastating blow the discovery dealt to their mathematics is not known, but they tried to keep the result secret and vengeance was taken on Hippassus[13] for his having divulged the secret when it finally got out. What is known is: i. the Pythagoreans discovered the irrational numbers; ii. Plato reports in the dialogue *Theætetus* that Theodorus proved the irrationality of the square roots of 3, 5, and "other examples" up to 17, suggesting the irrationality of $\sqrt{2}$ already known by Plato's time; and iii. Aristotle alludes to a proof of the irrationality of $\sqrt{2}$ in his *Prior Analytics*.

2.3 Theorem. $\sqrt{2}$ *is irrational.*

The proof that Aristotle hints at is assumed to be the familiar one: Suppose $\sqrt{2} = p/q$ is rational, with p/q reduced to lowest terms. Then $2q^2 = p^2$ and p must be even, say $p = 2r$. Then $2q^2 = (2r)^2 = 4r^2$, whence $q^2 = 2r^2$. Thus q is even, contradicting the assumption that p/q is reduced.

By the Pythagorean Theorem, $\sqrt{2}$ is the length of the diagonal of the square of side 1, and Theorem 2.3 thus gives the side and diagonal of a square as incommensurable line segments.

There is another, in some ways more natural, candidate for mankind's first irrational number. This is the *golden ratio*, revered for its self-duplication property— which property ensures its irrationality. Also known as the *divine proportion*, the golden ratio ϕ is the ratio of the length to width[14] of the sides

---

[12] Cf. footnote 11.

[13] It is not clear who meted out justice to Hippassus— his fellow Pythagoreans or the gods. I found four distinct stories in the history books.

[14] The letter "ϕ" is now standard for the golden ratio. What is not quite standard is the decision as to whether it represents the ratio of the long to short sides or the short to long ones. Thus, what I call ϕ another would call ϕ^{-1}.

of the rectangle $ABCD$ of *Figure 14* on the next page with the following property. If one rotates a short side, say BC, until it lies evenly on the long side and then deletes the resulting square, the rectangle $AEFD$ left over is similar to the original:

$$\frac{AB}{BC} = \frac{DA}{AE}. \tag{3}$$

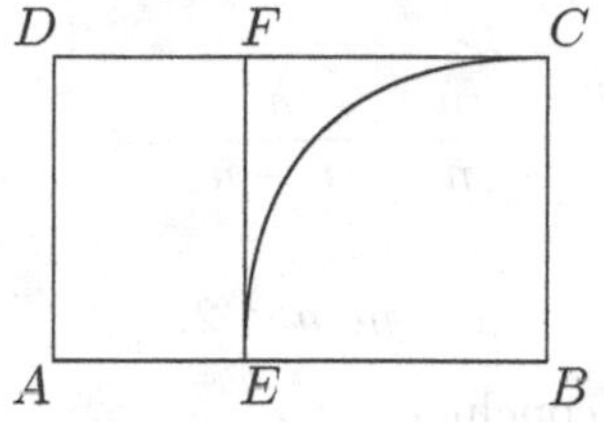

Figure 14

Choosing the short side to be 1, so that the long side AB is ϕ itself, this self-replication (3) reads

$$\frac{\phi}{1} = \frac{1}{\phi - 1}, \tag{4}$$

i.e. $\phi^2 - \phi - 1 = 0$, whence

$$\phi = \frac{1 \pm \sqrt{5}}{2}.$$

Since we do not allow negative lengths, this yields

$$\phi = \frac{1 + \sqrt{5}}{2} = 1.68\ldots \tag{5}$$

2.4 Theorem. ϕ *is irrational.*

Of course, this follows from the irrationality of $\sqrt{5}$, which can be established exactly as we established the irrationality of $\sqrt{2}$. However, there is an alternative proof that applies in these cases— and most naturally for ϕ. Geometrically, the proof runs as follows. Suppose ϕ were rational, i.e. AB and BC are commensurable with

$$\frac{AB}{BC} = \frac{m}{n}.$$

Lay BC off AB and delete the square obtaining $AEFD$ similar to the first rectangle. One can now subtract AE from AD to obtain a smaller rectangle still similar to the first. Obviously, this can go on forever. But, if AB and BC are measured by the same unit, so is $AE = AB - BC$. And so are $AD - AE$ and all the sides of the successive rectangles. Eventually, however, one of these will be shorter than the unit used to measure them. This is a contradiction, whence AB and BC are not commensurable.

I present the proof a bit more formally, as well as more arithmetically.

Proof of Theorem 2.4. Suppose ϕ were rational, i.e. suppose

$$\phi = \frac{m}{n} \tag{6}$$

for some positive integers $m > n$. Now (4) tells us

$$\frac{m}{n} = \frac{1}{\frac{m}{n} - 1},$$

i.e.

$$\frac{m}{n} = \frac{n}{m - n}. \tag{7}$$

By (5), we conclude

$$1 < m/n < 2, \tag{8}$$

from which we can further conclude

$$n < m < 2n \tag{9}$$

and

$$0 < m - n < n. \tag{10}$$

We now define two infinite sequences m_i, n_i (corresponding to the lengths of the long and short sides of the $(i+1)$-th rectangle in terms of the common unit for AB and BC). Start with m_0, n_0 being any pair of whole numbers satisfying (6). Given m_i, n_i satisfying (6), define $m_{i+1} = n_i, n_{i+1} = m_i - n_i$. By (10), n_{i+1} is not 0 and we can divide m_{i+1} by n_{i+1}. By (7),

$$\frac{m_{i+1}}{n_{i+1}} = \frac{m_i}{n_i} = \phi.$$

I.e., the process can be iterated infinitely often.

But, (9) tells us

$$m_{i+1} = n_i < m_i,$$

while (10) yields

$$0 < n_{i+1} = m_i - n_i < n_i.$$

Thus we have two infinite descending sequences,

$$m_0 > m_1 > \ldots \quad \text{and} \quad n_0 > n_1 > \ldots$$

of positive integers, which cannot be.   □

2.5 Remark. One can avoid the appeal to infinite descent by stipulating m, n to be a pair for which m or n is minimum satisfying (6) and then producing $n, m - n$ with $n < m$ by (9) and $m - n < n$ by (10) to get the contradiction. Another approach is to assume m/n to be in lowest terms and apply (7) to get $m^2 - mn = n^2$ and conclude that any prime divisor of n (respectively, m) must be a prime divisor of m (respectively, n). This forces, for m/n in lowest terms, $m = n = 1$, which choice does not yield ϕ. In the 1950s, the logician John Shepherdson proved that the irrationality of such numbers as

$\sqrt{2}$ and ϕ cannot be proven without the use of some variant of mathematical induction, be it the Method of Infinite Descent, the Least Number Principle, or some consequence thereof such as the ability to reduce fractions to lowest terms. That said, the three variants of the proof are worth noting. The appeal to infinite descent most closely mirrors the geometric approach; the appeal to the Least Number Principle clears away some of the grubbier details of the infinite descent; and the appeal to reduced fractions would be most palatable to students unfamiliar with induction.

One of the things making ϕ plausible as a candidate for the first irrational number is the geometric intuition behind the proof. Another is the familiarity of the Pythagoreans with the golden ratio. Indeed, they were intimately familiar with it. One of the symbols of the Pythagorean brotherhood was the pentagram or 5-pointed star which is obtained from the regular pentagon by connecting alternate vertices of the latter by lines. When one does this, the centre of the star is another pentagon, as in *Figure 15*, below.

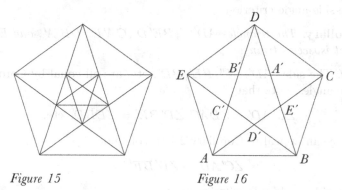

Figure 15   *Figure 16*

The uncluttered pentagram with its encompassing pentagon (*Figure 16*) is already fascinating to contemplate without the infinite series. It contains quite a few sets of congruent and similar triangles, equalities, and, hidden amongst its various proportions, more than one instance of ϕ. There is such a wealth of material here that it is easy to go astray in trying to demonstrate any of these facts. It is thus not the sort of thing to try to present to class without careful preparation[15]. Thus, let me outline some of its features leading up to the presence of ϕ.

2.6 Lemma. *The triangles ABC, BCD, CDE, DEA and EAB are congruent isosceles triangles.*

Proof. By regularity, $AB = BC = CD = DE = EA$ and $\angle ABC = \angle BCD = \angle CDE = \angle DEA = \angle EAB$. The familiar side-angle-side criterion for congruence yields the congruences. Repeating the first equation, $AB = BC$, shows ABC, and thus its congruent companions, to be isosceles. □

---

[15] This is the voice of experience speaking.

2.7 Corollary. *The triangles* $ACD, BDE, CEA, DAB,$ *and* EBC *are congruent isosceles triangles.*

Proof. By Lemma 2.6 and the side-side-side criterion.   □

2.8 Corollary. *The triangles* ABD', BCE', CDA', DEB' *and* EAC' *are congruent isosceles triangles.*

Proof. Consider ABD' and BCE'.

$$\angle D'AB = \angle CAB = \angle DBC = \angle E'BC$$

by Lemma 2.6. But

$$\angle BCE' = \angle BCA = \angle BAC = \angle CBD = \angle CBE'$$

by two applications of the Lemma. Similarly, $\angle BAD' = \angle ABD'$. Thus the triangles are isosceles with a common repeated angle. Moreover, the sides AB and BC between these angles are equal and we get congruence by appeal to the angle-side-angle criterion.   □

2.9 Corollary. *The triangles* $AD'C', BE'D', CA'E', DB'A',$ *and* $EC'B'$ *are congruent isosceles triangles.*

Proof. The sides AD', AC', BE', BD', etc. are all equal by Corollary 2.8. As to the angles, note that

$$\angle C'AD' = \angle DAC, \angle D'BE' = \angle EBD, \text{ etc.}$$

whence we can appeal to Corollary 2.7 to conclude

$$\angle C'AD' = \angle D'BE' = \ldots$$

Thus, the side-angle-side criterion applies.   □

We haven't exhausted all the congruent isosceles triangles in *Figure 16*. To proceed further with them, however, we seem to need a little lemma. To state it, let

$$\alpha = \angle ABC, \quad \beta = ABD', \quad \gamma = \angle C'AD'.$$

2.10 Lemma. $\gamma = \beta$.

Proof. Note that triangles $AD'C'$ and ACD, being isosceles with a shared non-repeating angle, are similar, whence

$$\angle AD'C' = \angle ACD = \beta + \gamma.$$

But also,

$$\angle AD'B = \pi - 2\angle ABD' = \pi - 2\beta,$$

whence

$$\pi = \angle AD'C' + \angle AD'B = \beta + \gamma + \pi - 2\beta = \gamma + \pi - \beta,$$

whence $\gamma = \beta$.   □

2.11 Corollary. *i. The triangles* ABE', BCA', CDB', DEC', *and* EAD' *are congruent isosceles triangles.*
ii. The triangles BAC', CBD', DCE', EDA', *and* AEB' *are congruent isosceles triangles.*
iii. The triangles of assertions i and ii are congruent to each other.

Proof. Note that

$$\angle AE'B = \pi - \angle CE'B$$
$$= \pi - (\pi - 2\beta) = 2\beta$$

and

$$\angle ABE' = \angle ABD' + \angle D'BE' = \beta + \gamma = 2\beta.$$

Thus ABE' is an isosceles triangle and, in particular,

$$AE' = AB. \tag{11}$$

The angle calculation holds for each of the triangles in question, whence they are all isosceles with repeated angle 2β. Moreover, since $AB = BC =$ etc., the repeated sides are all the same.   □

2.12 Corollary. *The triangles* ACB', BDC', CED', DAE', *and* EBA' *are congruent isosceles triangles and are congruent to the triangles of Lemma 2.6.*

I leave the proof to the reader. There is plenty more to explore— the parallelism of the lines AB and $A'B'$, or BC and $B'C'$, etc.; the congruences of parallelograms $ABCB'$ and others; and so on. The two facts needed beyond Corollary 2.11 for our purposes are the following two results.

2.13 Corollary. *The triangles of Corollaries 2.7, 2.9, and 2.11 are all similar.*

2.14 Corollary. *The pentagon* $A'B'C'D'E'$ *is regular.*

Corollary 2.13 follows from the observation that these triangles all have one angle equal to β and two equal to 2β. As for Corollary 2.14, note that the sides are equal to the bases of the congruent triangles of Corollary 2.9 and the interior angles (e.g. $\angle C'D'E'$) equal corresponding angles ($\angle BD'A$) of the congruent triangles of Corollary 2.8. Hence the sides and angles of $A'B'C'D'E'$ are equal.

I promised that we would find the golden ratio in this figure. It occurs several times.

2.15 Theorem.

i. $\quad \dfrac{AC}{AB} = \phi$

ii. $\dfrac{AB}{AD'} = \phi$

iii. $\dfrac{AD'}{D'E'} = \phi.$

Proof. i. By Corollary 2.13, ABD and $E'CD$ are similar. Now,

$$\begin{aligned}
\frac{AC}{AB} &= \frac{AD}{AB}, \quad \text{by Corollary 2.7} \\
&= \frac{DE'}{E'C}, \quad \text{by similarity} \\
&= \frac{DE'}{AC - AE'} \\
&= \frac{AE'}{AC - AE'}, \quad \text{by Corollary 2.11 or 2.12} \\
&= \frac{AB}{AC - AB}, \quad \text{by (11).}
\end{aligned}$$

ii. Triangles ABD and $AD'E$ are similar by Corollary 2.13, whence

$$\frac{AB}{AD'} = \frac{AD}{AE} = \frac{AC}{AB}.$$

[Alternatively, one can observe that $\angle BD'A = \angle C'D'E' = \alpha$, whence ABD' is similar to ACB and

$$\frac{AB}{AD'} = \frac{AC}{AB}.]$$

iii. Observe

$$\begin{aligned}
\frac{AB}{AD'} &= \frac{AD'}{AB - AD'}, \quad \text{by part ii} \\
&= \frac{AD'}{AE' - AD'}, \quad \text{by (11)} \\
&= \frac{AD'}{D'E'}.
\end{aligned} \qquad \square$$

2.16 Corollary. $A'D' = AD'.$

Proof. Applying part i of the Theorem to the small pentagon $A'B'C'D'E'$, we see

$$\frac{A'D'}{D'E'} = \phi.$$

But part iii tells us

$$\frac{AD'}{D'E'} = \phi.$$

The equality follows. $\square$

Assuming the Pythagoreans' interest in the pentagram predates the discovery of irrational numbers, they would have known all these facts about it. Assuming AB and AC commensurable, one could go on to note that the unit measuring AB and AC also measures $AD' = AC - AB, A'D' = AD'$, and $D'E' = AB - AD'$. That is, this unit also measures the corresponding segments of $A'B'C'D'E'$ and its inscribed pentagram. Hence it measures everything in the next pentagram, etc. In this way one produces an infinite sequence of smaller pentagons with sides all measured by a single common unit, which is impossible.

Thus we see that, whether contemplating the golden rectangle itself or delving into the deeper mysteries of the pentagram, the golden ratio ϕ could, on assumption of its rationality, easily lead to the realisation that something was wrong. The geometric representation of irrationality via the repeated production of ever smaller similar figures is appealing. The importance the Greeks attached to the golden ratio, as well as its multiple appearance in the pentagram add weight to the argument that ϕ was the first number recognised to be irrational. On the other hand, the primacy of arithmetic among the Pythagoreans prior to the discovery, the utter simplicity of the argument, and its mention in the earliest extant literature on the irrationals all support the argument for $\sqrt{2}$ as the first irrational.

3 The Eudoxian Response

Strictly speaking, the last section should have been titled "The Discovery of Incommensurables" rather than "The Discovery of Irrational Numbers". For, irrational numbers were not officially numbers in Greek mathematical ontology. For that matter, neither were rational numbers. Rational numbers were *ratios*, relations between whole numbers and, by extension, relations between line segments. Ratio in geometry had an operational definition. If one could use a segment to measure two other segments AB and CD by and the measures came out as 7 and 3, respectively, then the ratio of AD to CD was 7 to 3. If this unit were, say, a meter long and one replaced it by a ruler 10 centimeters long, the measures would now be 70 and 30— still a 7 to 3 ratio. But there is no such common measure for incommensurable segments, such as the diagonal and the side of a square. This raises a fundamental difficulty over and above the obvious invalidation of previously acceptable[16] proofs, possibly turning some accepted theorems into open problems, there is now the problem that one doesn't know what— if anything— the ratio of incommensurable segments is.

The eventual Greek solution to the problem by Eudoxus, a contemporary of Plato, would nowadays be recognised as a compromise, a stopgap that solved the problem at hand but in no way addressed the truly fundamental difficulty. This would finally be done in the 19th century by various means, in one case by completing the Eudoxian solution.

---

[16] I write "accept*able*" instead of "accept*ed*" because, of course, any specific statement about this or that "Pythagorean" proof is only hypothetical reconstruction.

The best way to explain the Eudoxian solution to the problem is to use modern terminology. The main thing one wants to do with proportions is to compare them. The comparisons are real numbers, comparison means determining equality or inequality, and we know how to do this for rational numbers. Eudoxus simply postulates the density of the rational numbers among the possible ratios, defining, for two proportions α/β and γ/δ,

$$\frac{\alpha}{\beta} = \frac{\gamma}{\delta} \quad \text{iff} \quad \text{for all positive integers } m, n,$$

$$
\begin{aligned}
\frac{m}{n} &< \frac{\alpha}{\beta} \quad \text{iff} \quad \frac{m}{n} < \frac{\gamma}{\delta} \text{ and} \\
\frac{m}{n} &= \frac{\alpha}{\beta} \quad \text{iff} \quad \frac{m}{n} = \frac{\gamma}{\delta} \text{ and} \\
\frac{m}{n} &> \frac{\alpha}{\beta} \quad \text{iff} \quad \frac{m}{n} > \frac{\gamma}{\delta}
\end{aligned}
$$

$$\text{iff} \quad \text{for all positive integers } m, n,$$

$$\left. \begin{aligned}
[m\beta &< n\alpha \quad \text{iff} \quad m\delta < n\gamma \text{ and} \\
m\beta &= n\alpha \quad \text{iff} \quad m\delta = n\gamma \text{ and} \\
m\beta &> n\alpha \quad \text{iff} \quad m\delta > n\gamma].
\end{aligned} \right\} \tag{12}$$

Eudoxus also added an axiom, now known as the *Archimedean Axiom* in honour of Archimedes, who made liberal use of it: given any segment AB, however large, and any segment CD, however small, some multiple of CD will be larger than AB.

Euclid repeats (12) as Definition V-5 in the *Elements*. He does not explicitly assume the Archimedean Axiom, preferring instead to define two magnitudes to *have a ratio* if each magnitude is capable upon multiplication of exceeding the other (Definition V-4). The Euclidean rôle of the Archimedean Axiom is primarily to guarantee that magnitudes have ratios and is largely unneeded by Euclid whose propositions largely concern magnitudes that are assumed to have ratios.[17] [18]

The assumption that magnitudes have ratios comes in to play in another manner.

---

[17] "Largely" is not the same as "always". Euclid implicitly assumes magnitudes to have ratios in some of his proofs (e.g., V-8, VI-2). Thus, Definition V-4 is often taken as asserting the Archimedean Axiom.

[18] I note in reading Victor Katz's textbook, there is another interpretation of Definition V-4: some types of magnitudes do not have ratios. He points to the angle between the circumference of a circle and a tangent to the circle as an example of a magnitude that has no ratio to, say, a rectangular angle. Of course, no 0 degree angle has such a ratio. Euclid did not consider this to be a 0 angle, however, as evidenced by his Proposition III-16 in which he shows it to be smaller than any proper rectilinear angle, but does not draw the obvious conclusion. Indeed, this angle, known as the *horn angle* or *contingency angle*, would puzzle mathematicians through the centuries.

3.1 Lemma. *Let $\alpha, \beta, \gamma, \delta$ be magnitudes.*

$$\frac{\alpha}{\beta} = \frac{\gamma}{\delta} \quad \textit{iff} \quad \textit{for all positive integers } m, n, \, [m\beta < n\alpha \textit{ iff } m\delta < n\gamma].$$

Proof. The left-to-right implication is immediate.
To prove the right-to-left implication, assume

$$\forall mn[m\beta < n\alpha \quad \text{iff} \quad m\delta < n\gamma]. \tag{13}$$

We have to show

$$\forall mn[m\beta = n\alpha \quad \text{iff} \quad m\delta = n\gamma] \tag{14}$$

$$\forall mn[m\beta > n\alpha \quad \text{iff} \quad m\delta > n\gamma]. \tag{15}$$

Toward proving (14), assume m, n given such that

$$m\beta = n\alpha. \tag{16}$$

If $m\delta \neq n\gamma$, then either $m\delta < n\gamma$ or $m\delta > n\gamma$. By (13), the first of these implies $m\beta < n\alpha$ and hence cannot hold. Assume, accordingly, that $m\delta > n\gamma$, so

$$m\delta - n\gamma > 0. \tag{17}$$

By (16), for any $k > 1, km\beta = kn\alpha$, whence

$$(km - 1)\beta < kn\alpha$$

and, by (13)

$$(km - 1)\delta < kn\gamma,$$

i.e.

$$km\delta - kn\gamma < \delta$$

i.e.

$$k(m\delta - n\gamma) < \delta.$$

But, applying the Archimedean Axiom to $m\delta - n\gamma$, the applicability of which is guaranteed by (17), for some $k, k(m\delta - n\gamma) > \delta$ and we have a contradiction. Thus

$$m\beta = n\alpha \quad \text{implies} \quad m\delta = n\gamma.$$

The symmetric argument yields the converse implication and we've established (14).

To establish (15), suppose

$$m\beta > n\alpha.$$

If $m\delta > n\gamma$ does not hold, then either

$$m\delta = n\gamma \text{ and it follows by (14) that } m\beta = n\alpha$$

or
$$m\delta < n\gamma \text{ and it follows by (13) that } m\beta < n\alpha.$$

Thus $m\delta$ must be greater than $n\gamma$. □

The practical significance of this Lemma is not that it cuts down on the number of conditions one must prove to verify two ratios equal, but that it guarantees trichotomy and allows one to prove two ratios equal by showing neither one to be greater than the other, where (Euclid, Definition V-7) one defines inequality by:

$$\frac{\alpha}{\beta} > \frac{\gamma}{\delta} \qquad \text{iff} \qquad \exists m,n[n\alpha > m\beta \text{ but } n\gamma \not> m\delta],$$

i.e., iff for some m, n,

$$\frac{\gamma}{\delta} < \frac{m}{n} < \frac{\alpha}{\beta}.$$

Note that

$$\neg\left(\frac{\alpha}{\beta} > \frac{\gamma}{\delta}\right) \qquad \text{iff} \qquad \forall mn[m\beta < n\alpha \text{ implies } m\delta < n\gamma]$$

and

$$\neg\left(\frac{\gamma}{\delta} > \frac{\alpha}{\beta}\right) \qquad \text{iff} \qquad \forall mn[m\delta < n\gamma \text{ implies } m\beta < n\alpha],$$

whence the two conditions taken together yield both implications of the condition of Lemma 3.1 for equality.

Following such basic definitions, Euclid devotes Book V of the *Elements* to developing the theory of proportions and begins Book VI with Theorem 2.1 and its extension to parallelograms. The proof is simple enough that its inclusion will not do much harm to my intention to supplement rather than overlap most textbooks. As before, we assume proven that two triangles of equal height and base have equal area. We also assume that, if two triangles have equal height but unequal bases, the one with the larger base is the one of greater area.

Proof of Theorem 2.1. Let ABC and DEF be as in *Figure 12* and let m, n be arbitrary positive integers. Extend the base of ABC by tacking on $m-1$ copies of AB, thereby obtaining a new triangle AB_mC as in *Figure 17*.

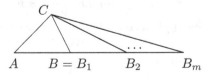

Figure 17

Clearly AB_mC has area $m \times ABC$. Similary, extend DE to obtain a triangle DE_nF of base $n \times DE$ and area $n \times DEF$. We have

$$m \times ABC < n \times DEF \quad \text{iff} \quad AB_mC < DE_nF$$
$$\text{iff} \quad AB_m < DE_n$$
$$\text{iff} \quad m \times AB < n \times DE.$$

Thus Lemma 3.1 tells us

$$\frac{ABC}{DEF} = \frac{AB}{DE}.$$ □

The appeal to Lemma 3.1 was not necessary here. We could successively replace all the $<$'s by $=$'s and then by $>$'s in the given proof and thereby obtain a proof using the original definition of equality of ratios. The real usefulness of Lemma 3.1 comes in the more complex proofs, e.g. in Euclid's determination of the area of a circle.

Euclid's proof that the ratio of the areas of two circles is the same as the ratios of the areas of the squares on their diameters is the most often presented example of a proof by the method of exhaustion. It is the Greek equivalent of a limit argument. Some textbooks present it in full and some, like Katz, just outline the proof. I prefer not to present it in too much detail, but we ought to discuss it. For, a point seldom made is that the proof is not without its weaknesses.

Euclid's Proposition on the area of a circle depends on three lemmas. First, there is the existence of a fourth proportional (Proposition VI-12).

3.2 Lemma. *For given magnitudes α, β, γ, there is a magnitude δ such that*

$$\frac{\alpha}{\beta} = \frac{\gamma}{\delta}.$$

Euclid proves this for line segments, for which it is an easy enough matter. Suppose for the sake of argument that $\alpha < \gamma$. Choose points A, B, C on a line segment with $AB = \alpha$, $AC = \gamma$; draw a perpendicular BD to AC at B so that $BD = \beta$; extend AD at least as far as AC and draw a perpendicular to AC at C. The point E on this perpendicular intersecting AD will be such that $CE = \delta$. For, the triangles ABD and ACE will be similar. (Cf. *Figure 18*.)

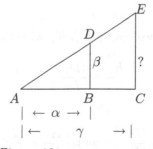

Figure 18

The result also holds for areas of rectilinear figures. For, Euclid proves any rectilinear figure equal in area to a rectangle and his proof of the Pythagorean

Theorem shows (as he also proves explicitly by a different argument in Book II) every rectangle is equivalent to a square. Thus, if one has three rectilinear magnitudes α, β, γ, one can assume them squares of sides α', β', γ', find the fourth proportional δ' of these and the area of the square of side δ' will be the fourth proportional δ of the given rectilinear magnitudes. He has not shown how to construct such a fourth proportional should one of the plane figures be a circle. This can be shown once one has proven the areas of the circles are to each other as the squares on their diameters, or one has shown how to construct a square equalling a given circle (squaring the circle), or one has added an axiom asserting the existence of a plane figure of any given magnitude.

Euclid's second lemma (Proposition X-1) is a sort of variant of the unassumed Archimedean Axiom dealing with small rather than large magnitudes.

3.3 Lemma. *Suppose α and β are magnitudes with α smaller than β. If one deletes from β more than half of it, deletes more than half the remainder from itself, and continues doing this, after finitely many steps one will get a magnitude smaller than α.*

Proof. Let AB and CD be line segments of magnitudes β and α, respectively, with $\alpha < \beta$. By the Archimedean Axiom for some number $n, n \times CD$ is greater than AB. On AB carry out the deletion $n - 1$ times. The smallest piece, say DE is less than CD. For, it is the smallest of n pieces totaling AB and $n \times CD$ is greater than AB.   □

Euclid's third lemma is the polygonal version of the desired result (Proposition XII-1).

3.4 Lemma. *Similar polygons inscribed in circles are to one another as the squares on the diameters of the circles.*

Armed with these three lemmas, Euclid is now ready to determine the area of a circle (Proposition XII-2).

3.5 Theorem. *The areas of two circles are to each other as the areas of the squares on their diameters.*

Proof sketch. Imagine two circles K_1, K_2 with diameters D_1, D_2, respectively. To show

$$\frac{K_1}{K_2} = \frac{(D_1)^2}{(D_2)^2},$$

it suffices to show

$$\frac{K_1}{K_2} \not> \frac{(D_1)^2}{(D_2)^2} \quad \text{and} \quad \frac{K_1}{K_2} \not< \frac{(D_1)^2}{(D_2)^2}.$$

Case 1. Suppose

$$\frac{K_1}{K_2} < \frac{(D_1)^2}{(D_2)^2}.$$

Let Σ be the fourth proportional to $(D_1)^2, (D_2)^2$, and K_1:

$$\frac{(D_1)^2}{(D_2)^2} = \frac{K_1}{\Sigma}.$$

Note that $\Sigma < K_2$.

Inscribe a sequence $P_1, P_2, \ldots$ of regular polygons inside K_2 as follows: P_1 is the inscribed square. P_{n+1} is obtained from P_n by doubling the number of sides: if A, B are successive vertices of P_n, bisect the arc AB of K_2 at a point C and replace the edge AB of P_n by the edges AC and CB. Note that in doing so, the excess $K_2 - P_{n+1}$ is less than half the excess $K_2 - P_n$. Hence Lemma 3.3 tells us, for some $n, K_2 - P_n < K_2 - \Sigma$. Choose such an n.

Inscribe a figure P'_n similar to P_n inside K_1. By Lemma 3.4, we have

$$\frac{P'_n}{P_n} = \frac{(D_1)^2}{(D_2)^2} = \frac{K_1}{\Sigma} > \frac{K_1}{P_n} > \frac{P'_n}{P_n},$$

a contradiction.

Case 2. Suppose

$$\frac{K_1}{K_2} > \frac{(D_1)^2}{(D_2)^2}.$$

Then

$$\frac{K_2}{K_1} < \frac{(D_2)^2}{(D_1)^2},$$

which case we have just proven impossible. □

The strengths and weaknesses of Euclid's *Elements* are made manifest in this proof. The strengths pretty much speak for themselves, but the weaknesses require some thought, and some class time should be devoted to them.

Euclid's *Elements* has often been viewed as the deductive system *par excellence*. His rigour and style served as model and inspiration for centuries. This is perhaps most dramatically demonstrated by Baruch Spinoza's emulation of the *Elements* in writing his book on Ethics. Yet one can question how important the deductive presentation was to Euclid. He did not cover all cases in his proofs, something he was certainly aware of as evidenced by the fact that, according to Proclus, many of the early commentators— lesser mathematicians than Euclid himself— devoted much of their energies to supplying the details in additional cases. For centuries, geometry was the theory of space, not merely a deductive system, and modern criticism of Euclid's logic is largely misdirected.

Modern criticism of Euclid began in the 19th century when Moritz Pasch added axioms for the notion of betweenness and proved that if a straight line entered a triangle it had to exit at a second point. This development reached its pinnacle with David Hilbert, who spent the 1890s teaching and reteaching the subject, his efforts culminating in 1899 with the publication of *Grundlagen der Geometrie* (*Foundations of Geometry*), a short booklet giving a highly axiomatic treatment of the subject. To Hilbert, the objects of study in an axiomatic treatment are *defined* implicitly by the axioms, and we are not allowed to use any properties of these objects not supplied by the axioms. Euclid did not follow the Hilbertian dictum, using evident properties of points, lines, and circles not supplied by his axioms. Hilbert too did not entirely succeed in this, but he went a long way and, indeed, as a deductive system, his axiomatisation was as close to perfection as had ever been seen. It would have made a lousy textbook however.

Now Euclid was not presenting a closed deductive system about objects defined, *à la* Hilbert, via their axioms. He was presenting geometry, a theory of space, deriving new results from old. I do not consider it a logical weakness that he would use additional obvious properties, such as the circle's possessing an interior and an exterior. Nor do I fault him for not axiomatising the notions of betweenness and order. Pure deductive systems have their uses, but pædagogy is not one of them and Euclid was writing the most successful textbook in the history of the world.

Even if one eschews the Hilbertian yardstick used by some authors to measure Euclid by[19], one can still find fault with Euclid. His treatment of magnitudes I find especially weak. The linear case he handles well. It is true that, in Proposition I-3, he assumes of two unequal line segments that one is greater than the other, but in Proposition I-2 he shows how a copy of one segment can be affixed to the endpoint of the other and one can then use the compass to place both segments on the same line and compare lengths using the Common Notion that the whole is greater than the part to establish trichotomy for magnitudes of line segments. This cannot be done directly for areas and with our modern knowledge of partial orderings we can imagine a geometry with incomparable areas.

The rectilinear case is unproblematic. He shows how, via decomposition or reference to parallelograms every polygonal figure is equal in area to a square, and the magnitudes of squares can be ordered by the magnitudes of their sides. And even without proving Theorem 3.5 it is clear that trichotomy for the magnitudes of areas of circles follows from the same property for the magnitudes of their diameters. But how do magnitudes of circles fit in with magnitudes of squares? And can one, *à la* Eudoxus, compare 7 copies of one circle with 3 of another? Without some axiom asserting the existence of, say, a square of any planar magnitude[20], I don't see how to make these comparisons in general.

---

[19] Cf. for example the discussion of Euclid given in David Burton's *The History of Mathematics*, the textbook used in my course.

[20] In terms of he notion of constructibility of the next chapter, all polygonal areas in the model of geometry given there are constructible numbers, while all circular areas are nonconstructible. Being real numbers, they are still linearly ordered, but the two sets are disjoint.

Again, without an axiom to the effect that all magnitudes were given by squares, I don't see how to effect the construction of the fourth proportional in the proof of Theorem 3.5 other than to assume the truth of the theorem being proven.

It could be that Euclid's reasoning is just circular. It could be that he just didn't see a problem because, in the back of his mind, he was thinking of magnitudes (upon designation of a unit) as numbers of some kind and the axiom I want was not necessary (to construct a square of area α, use a side of length $\sqrt{\alpha}$). Even so, his failure to address the issue is not the mere oversight that neglecting to discuss betweenness was. Betweenness was like the air, unnoticeable amidst so many more remarkable things. And, in any event, betweenness was not a problematic issue like number. While it may be true that "number" officially meant only $2, 3, 4, \ldots$ (or, maybe, $1, 2, 3, 4, \ldots$), the Greeks, particularly the astronomers, calculated freely with rational numbers in the form of sexigesimal fractions inherited from the Babylonians. They knew of numerical approximations to $\sqrt{2}$ and hence its numerical nature. The discovery that $\sqrt{2}$ is irrational raised serious foundational issues and caused a shift in emphasis from arithmetic to geometry, where it was clear what the diagonal of a square was even though one did not know what to make of $\sqrt{2}$.

Euclid's failure to come to grips with real numbers manifests itself in another nearly inexplicable way. Once he has proven

$$\frac{K_1}{K_2} = \frac{(D_1)^2}{(D_2)^2}$$

for circles K_1, K_2 of diameters D_1, D_2, respectively, it is only a tiny step to

$$\frac{K_1}{(D_1)^2} = \frac{K_2}{(D_2)^2}$$

and the mensuration formula $A = Cd^2$ for some constant C. Euclid would certainly have been aware of the existence of approximate values of π and could have used one of the inscribed regular polygons to give a rough lower estimate. Such estimates are not very good— a square yields $\pi > 2$ and an octagon only $\pi > 2.8$—, but it would have been a start. Such numerical work was only first undertaken by Archimedes, who also first proved the two definitions of π in terms of the circumference and the area to be identical, and then estimated π by the circumferences of inscribed and circumscribed regular polygons.

4 The Continuum from Zeno to Bradwardine

The discussion of the weaknesses of Euclid's treatment of magnitude in the last section, although it may seem somewhat tangential, really gets to the heart of the matter. With the Pythagoreans all had been number and geometry had been applied arithmetic. The discovery of irrationality changed all that. One did not know what to make of the new "numbers" or even if there were other stranger ones yet lurking in the shadows, but line segments were more-or-less tangible objects. Thus geometry became its own foundation, and even the foundation of the rest of mathematics. Indeed, after specifying a fixed unit, Euclid treated numbers as magnitudes and reduced number theory to geometry. Geometry, however, does not make a good foundation for mathematics.

One problem is the proliferation of types of magnitudes. Euclid is noncommittal on the identification of linear, planar, and solid magnitudes, i.e. the abstraction of a single notion of number shared by lengths, areas, and volumes. That he proves, for example, the existence of the fourth proportional for linear magnitudes and applies the result to planar magnitudes suggests them to be fundamentally the same. On the other hand, he does not apply the trisectibility of the line segment to conclude that of an angle, which suggests angular magnitudes to be possibly different from linear ones.[21]

Nongeometric magnitudes also occur in Greek mathematics. Archimedes discusses weight in several works, even to the point of considering ratios of weights. And, of course, there is that most mysterious of all magnitudes—time. In his book on *Mechanics*, late in Greek scientific history, Heron treated time as a geometric magnitude, even taking ratios, but, for the most part, time was treated more qualitatively than quantitatively by the Greeks.

I want to say that the Greeks, knowing mensuration, but not making it a formal part of their mathematics because it is arithmetic rather than geometric, had the notion of magnitude as number in the back of their minds and thus, if we ignore Euclid's agnostic stance on the Archimedean Axiom, we can interpret their magnitudes (upon specification of appropriate units) as real numbers with dimensions attached as in modern science. From the arithmetic standpoint we could then say that what they missed was a description of the real numbers in their totality, à la Dedekind. The situation, however, is more complicated than that. For, the switch from arithmetic to geometry was not just a change of language, but one of perspective. Geometric insight was appealed to and, in some crucial matters, we haven't any.

Geometry, as the Greeks inherited it from the Babylonians and Egyptians, dealt with mensuration. Its very name refers to land measurement. Under the Greeks it transformed into a theory of space. And, as "Nature abhors a vacuum", space was filled and geometry was a theory of matter, perhaps only

---

[21] And what should one make of his proof that the angle between the circumference of and a tangent to a circle is less than any given rectilinear angle without concluding it to be the null angle, as cited in footnote 18? He offers no similar potentially infinitesimal linear segment, also suggesting a different kind of magnitude. On the other hand, he did not explicitly assume the Archimedean Axiom for linear magnitudes either— perhaps because he assumed the two types of magnitudes to be similar(?).

of a special homogeneous matter like the æther, but matter nevertheless. In
the period between the discovery of irrationals and the Eudoxian response to
the problem, Democritus asked himself what happens if one takes an object
and keeps cutting it in half. His answer was that one could only do this so
many times before coming to a particle that couldn't be divided any further.
Such a particle he called an atom. Democritean atoms were *extended*— they
had dimension. Some believed one would arrive at a point, another form of
indivisible distinguished from the atom by its lack of extension, i.e. its lacking
in length, width, and breadth. Aristotle disagreed vehemently. To him, the
line did not consist of indivisibles, but was of one piece which could ever be
cut into shorter pieces. One can compare the lines of Democritus and Aristotle
by placing these lines under a very powerful microscope. That of Democritus
would look like a string of beads— indivisible particles lined one after another;
Aristotle's would look like the original line. But for the density, one could
imagine in like manner the line composed of points to be represented by a
dotted line which would look like a dotted line under the microscope.

Time was not geometry, space, or matter. Our one great temporal intuition
seems to be the separation of the past from the future by the present. The
present is an instant and instants are ordered. Whether an instant has duration
or not, it takes a moment to complete experiencing the present and we are
likely to think in terms of passage from one instant to the *next*, i.e. to think
of time as *discretely* ordered, like the integers. Geometry alone cannot tell
us whether the spatial and temporal lines should be the same. For this we
need motion, as when Aristotle[22] says, "If time is continuous, so is distance,
for in half the time a thing passes over half the distance, and, in general, in
the smaller time the smaller distance, for time and distance have the same
divisions".

One of Aristotle's goals in his *Physics* was to prove that the geometric
line was a continuous whole, capable of being subdivided any finite number
of times and thus *potentially* infinitely divisible, but not *actually* so. Another
goal was to refute the arguments of Zeno of Elea, although modern scholars
read the same goal into Zeno's argumentation.[23]

Zeno of Elea was born a century before Aristotle and is remembered today for his four paradoxes of motion which, on the surface, attempt to show
motion is impossible. The modern view is that he was actually attempting
to show motion impossible if one assumed space and time to be composed of
indivisibles, be they points or atoms. None of his work survives and our best
source of information on Zeno's paradoxes is Aristotle's brief statement of the
paradoxes and his several refutations thereof.

---

[22] *Physics*, Book VI. I am quoting from Florian Cajori, "The history of Zeno's arguments on motion: Phases in the development of the theory of limits", published
in ten parts in *American Mathematical Monthly* 22, 1915.

[23] Zeno's paradoxes are the subject of much philosophical literature. The best reference for the mathematician, however, is Cajori's series of articles cited in footnote
22, above.

The paradox known as *Dichotomy* goes as follows. To pass from point A to point Z one must first reach the midpoint B between A and Z. Then, to get from B to Z, one must first reach the midpoint C between B and Z. Etc. Thus, if space is made up of an infinity of points, to get from A to Z, one must visit infinitely many points in a finite amount of time, which is impossible. Hence one cannot get from A to Z and motion is impossible.

Nowadays, we dismiss this argument by appeal to the infinite series,

$$\frac{1}{2} + \frac{1}{4} + \frac{1}{8} + \ldots = 1, \tag{18}$$

applying it not only to the distance from A to Z but also to the time it takes to cover the distance. For, each successive step requires only half as long to complete as the previous one.

According to Aristotle, Zeno's argument is a fallacy. For one cannot actually subdivide an interval infinitely often. Infinite subdivision is only potential. Thus, Zeno's construction and argument cannot be carried out. This is not a particularly pleasing response, but it is in line with the modern explanation of (18) as the *limit* of finite sums and not an actual sum of infinitely many numbers.

A second paradox by Zeno called *Achilles* is almost identical to *Dichotomy*. The *Arrow* is aimed at refuting points. Consider an arrow in motion. At each instant, it "occupies its own space" and thus cannot be moving. Hence, at no instant is it moving, whence the moving arrow is at rest.

This is now explained away by observing that movement is not a property of an object that may or may not hold at a given instant, but is rather a difference in position of the object at distinct instants. Indeed, our definition of instantaneous velocity in the Calculus is given as the limit of the average velocities over progressively smaller intervals, which sounds very Aristotelian.

Aristotle dominated philosophy for nearly two millennia. He had been found compatible with Judaism, Islam, and, after some initial opposition, Christianity. Nevertheless, his views were not universally accepted nor his authority always accepted as final. Thus, for example, around the end of the 11th century or perhaps in the early years of the 12th, the arabic scholar al-Ghazzālī, known as Algazel in the west, wrote a summary of the views of ibn Sina (known as Avicenna in the west) and shored up Aristotle with no fewer than six refutations[24] of Democritean atoms, i.e. extended indivisibles. Before offering my favourite proof, let me quickly remind the reader that these atoms are the components of space, not merely matter placed in space.

Arrange 16 atoms in a 4×4 square as in *Figure 19* on the next page. Each side is composed of 4 equal parts. But the diagonal is also composed of 4. Hence the diagonal equals the sides, which is impossible because one proves in geometry that the diagonal of a square is greater than the sides.

---

[24] They can be found in section 52 on atomism in Grant's *Source Book in Mediæval Science* cited in the bibliography.

This proof just seems wrongheaded. Why must atoms be squares and not, say, hexagons? Because he demands no vacuum, the 4 × 4 arrangement of hexagons form a jagged honeycomb arrangement as in *Figure 20*. Now this "square" has three diagonals: 11-22-23-33-44 and 11-22-32-33-44 of length 5 and 41-42-32-23-13-14 of length 6. All of these are greater than the sides and al-Ghazzālī's argument fails for these atoms.[25]

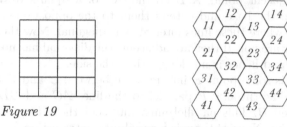

Figure 19

Figure 20

4.1 Remark. I cannot help but note that the ratios 5/4 and 6/4, though neither equal to $\sqrt{2}$, approximate it from below and above, respectively. Averaging these values gives

$$\frac{2 \cdot 5 + 6}{3 \cdot 4} = \frac{16}{3 \cdot 4} = \frac{4}{3}.$$

Now $(4/3)^2 = 16/9 = 1.7\overline{7}$, which is a bit smaller than 2. So far this does nothing for us[26]. But, if we square again, we get

$$\left(\left(\frac{4}{3} \right)^2 \right)^2 = \frac{16^2}{9^2} = 4 \times \left(\frac{8}{9} \right)^2,$$

the Egyptian value of π. If you wish to test your students' gullibility, you might point this out to them with a comment that this clears up the mystery of how the Egyptians arrived at this value.

The 6th proof is rather more convincing. When a wooden or stone wheel revolves, the parts near the centre move less than the parts near the rim because a circle near the centre is smaller than a circle near the rim. Now when the circle near the rim moves only one atom, the circle near the center will move less than an atom, thus dividing the atom, which is impossible. Against this we could argue that the rigidity of the wooden or stone wheel is illusory and that a smooth rotation at the rim might actually be accompanied by jerky rotation near the centre.

---

[25] But all is not lost— each of these diagonals is commensurable with the sides of the "square", a contradiction.

[26] Interestingly enough, if one doubles the weight of the singularly occurring diagonal 6/4 and gives a weight of 1 to the duplicate diagonal 5/4, one gets $(5+2\cdot6)/(3\cdot4) = 1.41\overline{6}$, which is a good approximation to $\sqrt{2}$. Indeed, its square is $2.0069\overline{4}$.

Moving on a couple of centuries we find ourselves with the Merton scholars in Oxford, and, in particular, with John Duns Scotus, who also argued that the continuum is not composed of indivisibles— specifically, of *adjacent* indivisibles. Aside from one repetition of al-Ghazzālī's claim that indivisibles commit one to equating the diagonal and the side, his arguments are more involved. One of them proceeds as follows[27].

Take two adjacent points A, B on the side of a square, draw lines from them parallel to the base and extend them to the opposite sides. Consider the points C, D where these lines intersect the diagonal. Now these are either adjacent or they are not. If they are adjacent, the diagonal has no more points than the side[28] and hence is not longer than the side.

Thus, there is some point E intermediate between C and D on the diagonal. Extend a line through E parallel to the lines AC and BD to the side containing A and B. This parallel must intercept the side at A, B or some intermediate point. None of these is a possibility: intersection at A makes AC and AE intersecting parallel lines; intersection at B does the same for BD and BE; and intersection at a point intermediate between A and B contradicts the assumption that A and B were adjacent.

I will grant that the construction of the square and parallel lines is probably not any more complicated than the construction of the midpoint using ruler and compass, but the construction of the midpoint had already been effected by Euclid (Proposition I-10), whose authority Duns Scotus seems to accept as gospel. The argument accompanying his construction, however, is more complicated than a simple appeal to the Euclidean construction and I conclude the latter didn't occur to him: his grasp of geometry was very weak.

Note that the proof just given was not aimed at Democritus, whose indivisibles were extended atoms, but probably at Henry of Harclay, "the first thorough-going atomist and... adherent of the existence of an actual infinite, in the later Middle Ages" according to Edward Grant, who includes an account of Harclay's views in his *Source Book* cited in footnote 24. Harclay believed in a continuum composed of infinitely many contiguous points.

And this brings us to Thomas Bradwardine, 14th century England's best mathematical mind, composing a treatise, the *Tractatus de Continuo* at Oxford. Bradwardine believed, with Aristotle, al-Ghazzālī, and most 14th century scholars, that the continuum did not consist of atoms or points but was a single entity capable of subdivision without end. However, he knew of unbelievers like Henry of Harclay. Indeed, in the *Continuo*, Bradwardine lists no fewer than five views on the composition of the continuum. It might be composed of

---

[27] Cf. Grant's book cited in footnote 24 for more details and another of his arguments.

[28] Logic must not have been Scotus's strong point. He should argue that either the points C, D are always adjacent and thus the diagonal has no more points than the side, or that, for some pair A, B, the points C, D are not adjacent.

1. a single piece divisible without end (Aristotle, al-Ghazzālī, Bradwardine and most of his contemporaries)

2. corporeal indivisibles, i.e. atoms (Democritus)

3. finitely many points (Pythagoras, Plato, Waltherus Modernus)

4. infinitely many adjacent points (Henricus Modernus = Henry of Harclay)

5. infinitely many densely packed points (Robert Grosseteste).

This treatise is, apparently, not the least bit important in the history of mathematics. As of 1987[29], only three surviving copies were known to exist—only one of them complete—, and only one oblique reference to the treatise in any other 14th century work had been found. Nonetheless, the *Continuo* is a valuable object for the student of mathematical history[30]. There are several

---

[29] Cf. John E. Murdoch, "Thomas Bradwardine: mathematics and continuity in the fourteenth century", in: Edward Grant and John E. Murdoch, eds., *Mathematics and Its Applications to Science and Natural Philosophy in the Middle Ages. Essays in Honor of Marshall Clagett*, Cambridge University Press, Cambridge, 1987

[30] Or, it would be if it were readily available in English translation. Briefly, the modern publication history of the *Continuo* is this. In 1868, Maximilian Curtze reported on the two then known codices of the treatise, the complete one in Thorn (now: Toruń) and the partial one in Erfurt. In 1936 Edward Stamm published a more detailed description of the Thorn manuscript, "Tractatus de continuo von Thomas Bradwardina", *Isis* 26 (1936), pp. 13 - 32. However, Stamm's article has an extensive series of untranslated Latin quotes. In her book, *Die Vorläufer Galileis im 14. Jahrhundert (The Precursors of Galileo in the 14th Century)* published in Rome in 1949, Anneliese Maier made an apparently brief mention of the *Continuo*. Her work is apparently important for its identification of Waltherus Modernus and Henricus Modernus and does not cover the contents too fully. John Murdoch's unpublished dissertation, "Geometry and the Continuum in the Four-teenth Century: A Philosophical Analysis of Thomas Bradwardine's Tractatus de Continuo" (University of Wisconsin, 1957) gives a fuller treatment and includes the complete text of the *Continuo*. Another detailed treatment— published, but in Russian with a Latin appendix— is V.P. Zoubov, "Traktat Bradwardina O Kontinuume", *Istoriko-matematicheskiie Issledovaniia* 13 (1960), pp. 385 - 440. Zoubov is reported to give a detailed analysis of the treatise and the appendix covers the definitions, suppositions, and conclusions. A selection of definitions, suppositions, and conclusions appear in English in Marshall Clagett's *Science of Mechanics in the Middle Ages*, University of Wisconsin Press, Madison, 1959, and a single short statement from the *Continuo* was translated into English in Grant's *Source Book* (1974). Murdoch's account of Bradwardine in the *Dictionary of Scientific Biography* includes a brief, nontechnical description of the *Continuo*. Finally, for the English reader, there is Murdoch's 1987 paper cited in the pre-ceding footnote. This work gives a mathematically nontechnical summary of the contents of the *Continuo* and provides a corrected and more complete listing in

reasons for this. The axiomatic framing of the *Continuo*, as well as that of Bradwardine's other works, the *Geometria speculativa*, the *Tractatus de proportionibus velocitatum in motibus*, the *De incipit et desinit*, and even a religious tract *De causa Dei contra Pelagium*, illustrates the high regard Euclid's *Elements* was held in during the Middle Ages. Moreover, Bradwardine's treatment of the problem of the continuum is as thorough-going as one could hope to find and offers a good summary of the various views held during the early 14th century, if not much of a taste of the scholarly debate that was raging on the subject.[31]

As I say, Bradwardine mimics Euclid in presenting the *Continuo* in an axiomatic framework. He begins with a series of 24 definitions— of continua, lines points, etc.,— follows these with a series of 10 suppositions, and then proceeds to draw 151 conclusions. Unlike Euclid, however, the goal of his work is not to derive true conclusions from these definitions and suppositions, but to draw absurd conclusions from various hypotheses on the composition of the continuum out of indivisibles. In this, he is probably better compared to Girolamo Saccheri, who attempted to demonstrate the parallel postulate by assuming its negation and deriving absurdities therefrom.[32]

Bradwardine's conclusions are assertions of the form "If the continuum consists of indivisibles of such and such a type, then..." I would love to give a sample of one of these conclusions and its proof, but I cannot quite do it. The conclusions have been published—in Latin[33]—, but the proofs apparently haven't. There is one exception, but I can't say I understand the proof and can only offer a sort of reconstruction of it. Toward's this end, let me first state a lemma. Conclusion 38 of *de Continuo* reads (in paraphrase):

4.2 Lemma. *If continua consist of adjacent points, then each point in the plane has only 8 immediate neighbours.*

4.3 Corollary. *Under the same hypothesis, each point in 3-dimensional space has no more than 26 immediate neighbours.*

I have no idea how he proves these, but one can see how he could conclude the Lemma by imagining the "points" to be shaped like squares and to tile the plane with them as in *Figure 19*, where each interior square borders on exactly 8 others, sharing edges with 4 and vertices with 4 others. In the three dimensional case, think of the points as being cubic. Each cube has 8

Latin of the definitions, suppositions, and conclusions of the treatise as an appendix. My account is based on the paper of Stamm and the last cited one of Murdoch.

[31] Murdoch, *op. cit.*

[32] Saccheri, however, succeeded in forming the beginnings of non-Euclidean geometry; it is not clear what Bradwardine's positive accomplishment, if any, was.

[33] Cf. Murdoch's paper cited in footnote 30.

neighbours in its own plane, and 9 in each of the planes immediately above and below this plane, thus $9 + 9 + 8 = 26$ in all.[34][35]

Bradwardine applies Conclusion 38, i.e. Lemma 4.2, in establishing Conclusion 40, which I paraphrase as follows:

4.4 Theorem. *Under the same assumption, the right angle is the minimum angle and is not acute, and all obtuse angles are equal to each other.*

The proof as quoted in Stamm's paper cited in footnote 30 is as follows. "Let AB be a straight line with point C intermediate between A and B, and let DE be perpendicular to AB at C. In the immediate vicinity of C there are no more than 8 points, and 4 of these are on the lines AC, BC, DC, EC. The remainder would only number 4 and are in those 4 angles, therefore in whatever direction you choose there is only one, but an angle of less than one point does not exist nor is one point less than another by an earlier conclusion." This is as much of the proof as is given by Stamm and I confess not to be completely confident in my translation[36].

Look at *Figure 21*. Now CK does not define a straight line because C and K are not adjacent and continua are assumed to consist of adjacent points[37].

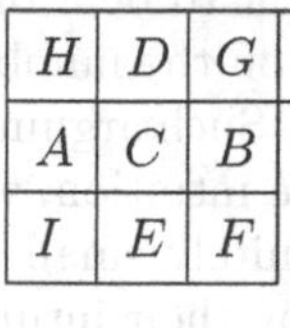

H	D	G	K
A	C	B	
I	E	F	

Figure 21

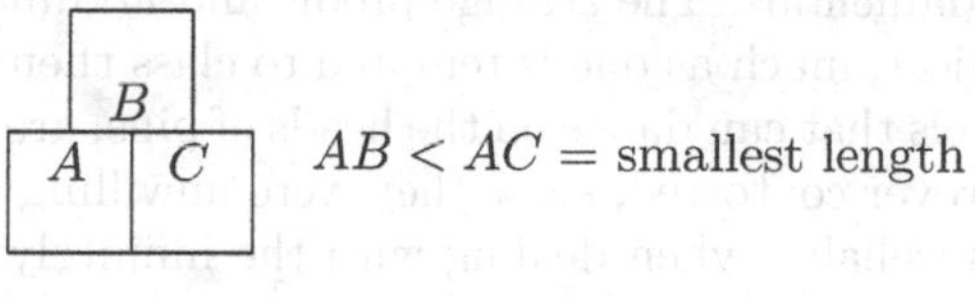

$AB < AC$ = smallest length

Figure 22

Therefore, the only lines making angles with CB at C are $CB, CG, CD, CH, CA, CI, CE$, and CF. The smallest nonzero such angles would be GCB and FCB, neither of which is right. Perhaps Bradwardine is arguing against the existence of the points F, G, H, I— he does use the subjunctive *sint* in first referring to them (*Igitur reliqua, tantum sint 4,...*).

The "earlier conclusion" referred to is probably some variant of the argument that nothing is shorter than a point. Think in terms of atoms, say

---

[34] Of course, why a dimensionless point should be thought of as having a shape at all is unclear. Further, one might expect these results to apply to Democritean atoms rather than to points. However, it is Henry of Harclay's adjacent points that Bradwardine wishes to refute with these particular conclusions. In any event, I cannot offer these squares and cubes as anything more than a heuristic. But see the next section of this Chapter.

[35] Note that this argument assumes the squares aligned in a checkerboard pattern, and not staggered like bricks. Staggering violates indivisibility: the length of the overlap of staggered blocks is less than the length of a side, but (the side of) a single point is the smallest distance.

[36] I know no Latin and am relying on a dictionary and the first chapter of my newly purchased Latin grammar.

[37] Note that CB extends to include a point directly below K and CG to include one directly above K, so neither B nor G is in line with C and K.

squares tiling the plane. (Cf. *Figure 22.*) Staggered squares cannot exist because the length of an overlap is less than an atom in size and as space is atomic, that length would have to be made up of atoms.

If one thinks of $BGDHAIEF$ as the circle of radius 1 point around C, one could possibly argue against the existence of G by noting that the horizontal component CB of the distance CG is less than CG itself, which is equal to CB (both being 2 points long). If one ignores C, this horizontal component is less than a point in size, contrary to the "earlier conclusion". One would conclude F, G, H, and I not to exist and the smallest angles around C would be DCB, ACD, ECA, and BCE— right angles all.

What should we learn from al-Ghazzālī, Duns Scotus, or Bradwardine? Mathematically, there is the obvious lesson that Euclidean geometry is not compatible with extended atoms or adjacent points. Historically, we can glean the strengths and weaknesses of these mathematicians. While certainly less skillful than the Greeks, they were not without some talent. Their interests were different. The Merton scholars were noted for their study of motion and the invention of the *latitude of forms*, a precursor to Descartes. The debate on the nature of the continuum and atomism demonstrates their interest in fundamentals. The strange proofs and argumentation brought to bear on the subject, much as one is tempted to class them with debates on the number of angels that can dance on the heads of pins, are to their credit. Such arguments, however contorted, show they were unwilling to rely on mere intuition, which is unreliable when dealing with the infinitely large and infinitely small. But, too, their arguments show precisely how they were misled by their intuitions whenever they applied their experience with the finite to the infinitely large and infinitely small. Arguments that a continuum does not consist of a dense set of points because any two line segments have the same number of points and must therefore have the same length offer a prime example of misapplying our experience with finite sets to infinite ones. And Lemma 4.2 is a nice example of applying our finite intuition to the infinitely small. In the next section I propose to take a closer look at this Lemma.

5 Tiling the Plane

Today we have no difficulty accepting points having no dimension. The 14th century scholars seemed at least unconsciously to have endowed these points with features of dimensional objects. Some of the arguments against successive points (i.e., Henry of Harclay), look like they were aimed at Democritean atoms. Let us assume this to be the case with Lemma 4.2.

Basic concepts of the macroworld like shape and size might have no meaning in the microworld. After all, a square is a configuration of lines which are themselves made up of atoms. Size is a concept abstracted from the process of measuring by comparing a longer to a shorter. Atoms in space may not have definite shapes or sizes. Let us assume, however, that they do.

Now, matter might be a mixture of different kinds of atoms of different sizes and shapes. This is particularly true if one postulates the different types of matter to be determined by these varying sizes and shapes. The finest of these particles would be the æther particles that fill space. On general principles, space itself devoid of other matter is homogeneous. This should mean all the atoms of pure space, i.e. of geometry, look alike; stated more geometrically, they are congruent.

Because there is no vacuum, the particles of space fill it up. Moreover, where they press up against each other, they must be flattened. We could argue for this by appeal to symmetry: whatever forces one particle to jut into another would be met by equal pressure in the other direction forcing the second particle to jut into the first. We may as well assume convexity while we are at it. Thus we imagine space to be filled with congruent, convex polygons if we are thinking of the plane, and congruent, convex polyhedra if we are thinking of 3-dimensional space.

On general symmetry considerations, we may as well assume the polygons (respectively, polyhedra) to be *regular*: all of the sides (respectively, faces) are equal. On making all of these simplifying assumptions, Lemma 4.2 reduces to examining the tilings of the plane by congruent regular polygons and its supposed corollary reduces to considering how 3-dimensional space can be filled by congruent regular polyhedra.

The following result was known already to Euclid.

5.1 Theorem. *Suppose the plane can be tiled by congruent regular n-gons. Then: $n = 3, 4, or\ 6$.*

Proof. Suppose the plane to be tiled by congruent regular n-gons and consider a vertex V of one of the polygons P of the tiling. Consider too the disposition of V relative to a second polygon P' on which V lies. V could be a vertex of P' as in *Figures 19* or *20*, or it could lie on one of the sides of P', other than a vertex, as in the brick pattern of *Figure 22*.

V cannot properly lie on the side of a third polygon P'' as well as on the side of P', because then the total angle around V would sum to greater than $360° - 180°$ for each of the straight angles inside P' and P'' plus the interior angle of P. Thus, V is either a vertex of all the polygons it lies on, or it lies properly on the side of one polygon between its vertices.

The interior angle of the regular n-gon is $180(n-2)/n$ degrees. For, the sum of the interior angles of an n-gon is $180(n-2)$ [38]. For $n > 4$,

$$\frac{1}{2} < \frac{n-2}{n} < 1,$$

and the interior angle is obtuse. Thus, if one has a tiling and V lies properly on the side of P', one cannot have $n > 4$ as the sum of the angles around V

---

[38] This is a simple induction on n.

is at least the sum of the interior angle of P, the straight angle of P', and another interior angle, yielding a total greater than $360°$.

Finally, if V is a vertex of all the polygons meeting at it, and there are k such polygons, the sum of the angles around V must be

$$k\frac{180(n-2)}{n} = 360, \tag{19}$$

i.e., we have

$$k = \frac{360n}{180(n-2)} = \frac{2n}{n-2},$$

and we conclude $n-2$ to be a divisor of $2n$. One readily checks this to be the case for $n = 3, 4, 6$ and not to be the case for $n = 5$.

To rule out $n > 6$, let

$$f(x) = \frac{2x}{x-2},$$

and note that

$$f'(x) = \frac{2(x-2)-2x}{(x-2)^2} = \frac{-4}{(x-2)^2} < 0,$$

for $x > 2$. Thus, f is strictly decreasing[39] and, as $f(6) = 3$, for $k = f(n)$ to be integral for $n > 6$ we must have $f(n) = 2$ or $f(n) = 1$. But, for $n > 2$, we have

$$f(n) = \frac{2n}{n-2} > \frac{2n}{n} = 2. \qquad \square$$

To establish Lemma 4.2, we now look at the possible tilings of the plane by equilateral triangles, squares, and regular hexagons. Ignoring the staggering, there are only three possibilities. The triangular tiling looks like *Figure 23*, below, the square tiling like *Figures 19* and *21*, and the hexagonal tiling like the honeycomb pattern of *Figure 20*. Now each hexagon has exactly 6 immediate neighbours and, counting intersections in a vertex, each square has exactly 8. If we look at *Figure 23* on the next page, however, each triangle has 12 immediate neighbours.

As I say, I know no Latin, so the "only 8" of Lemma 4.2 might actually mean "at most 8", in which case squares and hexagons are ok. But the triangles are still a problem. I would like to rule them out. We might try the following. Consider the pattern of *Figure 24*.

Is this a straight line composed of equilateral triangular atoms? It certainly looks it, but consider: B and C are both immediate successors to A on the right, and A has two more such neighbours on the left. These are too many

---

[39] If one wishes to avoid appeal to the Calculus, one calculates

$$f(n+1) - f(n) = \ldots = \frac{-4}{(n-1)(n-2)} < 0$$

for $n > 2$.

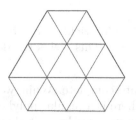

Figure 23   Figure 24

and I feel like declaring the triangles an unsuitable shape for geometric atoms. Besides, the fact that A and B jut into each other, one on the top and one on the bottom seems to me to destroy homogeneity: the line is now composed of two types of atoms— top-heavy and bottom-heavy.

The third dimension solves all of our problems. Euclid modified the proof of Theorem 5.1 to show there to be only 5 Platonic solids.

5.2 Theorem. *The only regular polyhedra are*
i. the tetrahedron (4 equilateral triangular faces)
ii. the cube (6 square faces)
iii. the octahedron (8 equilateral triangular faces)
iv. the dodecahedron (12 pentagonal faces)
v. the icosahedron (20 equilateral triangular faces).

Proof. As in the proof of Theorem 5.1, let n be the number of sides of the polygons forming the faces of the regular polyhedron. Further, let k denote the number of faces that come together at a vertex. As before, the interior angle of a corner of the polygon is $180(n-2)/n$ degrees and the sum of the angles around a vertex is k times this. If this sum equals 360, the faces are co-planar and we are tiling the plane. A sum less than 360 is possible, but a sum greater than 360 requires some folding and the polyhedron would not be convex. Thus, we must have

$$k\frac{180(n-2)}{n} < 360,$$

generalising (19) slightly. Thus,

$$k < \frac{2n}{n-2}. \tag{20}$$

Now, n must be at least 3 as a polygon has at least 3 sides. And k must be at least 3 as two faces meet in an edge, not a vertex. Moreover, for $n \geq 3$, $f(n) = 2n/(n-2)$ is strictly decreasing with $f(6) = 3$. this means that the only possible values of n are $n = 3, 4, 5$.

For $n = 3$, (20) yields $k < 6$, i.e. $k = 3, 4$, or 5. All these possibilities occur:
   $k = 3$: tetrahedron

$k = 4$: octahedron

$k = 5$: icosahedron.

For $n = 4$, (20) yields $k < 4$, i.e. $k = 3$. This yields the cube.

For $n = 5$, (20) yields $k < 10/3$, i.e. $k = 3$. This yields the dodecahedron.

$\square$

5.3 Remark. The above proof is not quite complete. While we have proven there to be only 5 pairs (n, k) for which there is a regular polyhedron consisting of n-gons, k of which meet in each vertex, we haven't shown this to be possible in each case in only one way. The best way to convince oneself of the truth of this is to take some construction paper, cut out a bunch of regular polygons, and start taping them together. If you have no options along the way, uniqueness follows.

5.4 Remark. Some authors prefer a more modern proof. Hans Rademacher and Otto Toeplitz[40], as well as Richard Courant and Herbert Robbins[41], prove Theorem 5.2 by appeal to Euler's formula[42][43],

$$2 = F + V - E, \tag{21}$$

relating the numbers of faces (F), edges (E), and vertices (V) of a convex polyhedron. Once one has (21), this is not too difficult. One starts by noting that, if one has a regular polyhedron composed of n-gons with k of them meeting at each vertex, then

$$2E = kV, \tag{22}$$

for, each edge contains exactly 2 vertices and, in counting the k edges associated with the V vertices, each edge gets counted twice. Similarly,

$$2E = nF, \tag{23}$$

as each edge lies on exactly 2 faces and counting the n edges associated with each of the F faces again counts each edge twice.

Rewriting (21) in terms of E, one has

$$2 = \frac{2E}{n} + \frac{2E}{k} - E = E\left(\frac{2}{n} + \frac{2}{k} - 1\right).$$

Because $2, E$ are positive, we must have

---

[40] *Von Zahlen und Figuren*, Springer-Verlag, Berlin, 1930.

[41] *What is Mathematics?*, Oxford University Press, Oxford, 1941.

[42] An equivalent statement was first conjectured by Rene Descartes in the 17th century, but was lost until the mid-19th century. Euler stated the formula without proof in the 1750s. The first generally accepted proof was given by Cauchy in the early 1800s. Euler's formula has an involved history that was studied and turned into an interesting philosophical dialogue by Imre Lakatos in his *Proofs and Refutations*, Cambridge University Press, Cambridge, 1976.

[43] Those not interested in the alternate proof should skip ahead to page 82.

$$\frac{2}{n} + \frac{2}{k} - 1 > 0. \tag{24}$$

As before, $n, k \geq 3$, whence

$$\frac{2}{n}, \frac{2}{k} \leq \frac{2}{3}.$$

Thus, if, according to (24), we have

$$\frac{2}{n} + \frac{2}{k} > 1, \tag{25}$$

we must have

$$\frac{2}{3} + \frac{2}{m} \geq \frac{2}{n} + \frac{2}{k} > 1$$

$$\frac{2}{m} > \frac{1}{3}$$

$$\frac{1}{m} > \frac{1}{6},$$

and $m < 6$, where m is either n or k. Thus, the only possibilities for (n, k) are the pairs (3,3), (3,4), (3,5), (4,3), (4,4), (4,5), (5,3), (5,4), and (5,5). Of these, (4,4), (4,5), (5,4), and (5,5) can be ruled out as

$$\frac{2}{5} + \frac{2}{5} < \frac{2}{4} + \frac{2}{5} = \frac{2}{5} + \frac{2}{4} < \frac{2}{4} + \frac{2}{4} = 1,$$

contrary to (25). Thus, modulo the proof of (21), we have an alternate proof of the existence of only 5 pairs (n, k) yielding regular polyhedra.

If we now replace (23) by

$$E = \frac{n}{2}F \tag{26}$$

and combine this with (22) to get

$$V = \frac{2E}{k} = \frac{2n}{2k}F = \frac{n}{k}F, \tag{27}$$

we can rewrite (21) as

$$2 = F + \frac{n}{k}F - \frac{n}{2}F$$

$$= F\left(1 + \frac{n}{k} - \frac{n}{2}\right) = F\left(\frac{2k + 2n - kn}{2k}\right)$$

and conclude

$$F = \frac{4k}{2k + 2n - kn}.$$

From this we can quickly read off the number of faces of the regular polyhedra. For example, $(n, k) = (3, 3)$ yields

$$F = \frac{12}{6 + 6 - 9} = \frac{12}{3} = 4,$$

the tetrahedron. From (26) and (27) one can proceed to determine the numbers of edges and vertices as well.

By whatever means, let us assume Theorem 5.2 completely proven and ask the next question: which regular polyhedra can "tile" 3-dimensional space. I.e., which of the 5 platonic solids can fill all of 3-dimensional space assuming all polyhedra of the same size?

5.5 Theorem. *Space cannot be filled with congruent copies of any regular tetrahedron, octahedron, dodecahedron, or icosahedron. It can be filled with cubes of equal size.*

The key to the proof is the *dihedral angle*, the angle between two adjacent faces of a regular polyhedron. One supposes space to be filled with copies of one of the platonic solids and considers an edge of one of these copies. Some number m of polyhedra must meet in the edge, and these polyhedra must fill the space around the edge, i.e. if δ is the dihedral angle, then $m\delta = 360$, i.e. δ is of the form $360/m$ for a whole number m. We can check when this is possible by calculating δ for the various regular polyhedra. This is a matter of mere computation using analytic geometry and trigonometry. Or, we can forgo the computation entirely by looking the angles up in a table:

Tetrahedron	70.53°
Cube	90°
Octahedron	109.47°
Dodecahedron	116.56°
Icosahedron	138.19°

One sees by examination that only the cube has a dihedral angle an integral multiple of which is 360°, whence only the cube can be used to fill space. The square tiling of the plane matches the cubic tiling of space. The hexagonal tiling of the plane has no 3-dimensional match, and although there are three platonic solids with equilateral triangular faces, none of them fills space to complement the triangular tiling of the plane. Thus, the image of cubic atoms filling space and square ones filling the plane is geometrically natural and there is at least an heuristic basis behind Bradwardine's mysterious Conclusion 38 (Lemma 4.2.)

With the results of the present section, I have changed direction slightly. I was trying to make the point that the switch from arithmetic to geometry, though it provided temporary stability in the face of irrationals, could not serve as a foundation for mathematics because our intuition is not sufficient to decide basic questions about the nature of geometry. The 14th century debate on this issue demonstrates this as convincingly as, and possibly more entertainingly than, the later troubles with infinitesimals. In any event, on the theory that students go through the same stages in understanding that a field went through historically, the difficulties the mediæval mathematicians had in accepting points may be of some interest to potential high school geometry teachers. Additionally, the fact that such strange looking results as Lemma 4.2 may have something behind them ought also to be of some interest.

6 Bradwardine Revisited

After printing and proofing the preceding sections, I received two items of interest in the mail that require some comment here. The first of these was a xerox of several pages devoted to Bradwardine from the German translation of A.P. Youschkevitch's book on the history of mathematics in the middle ages, sent to me by my friend and colleague Eckart Menzler-Trott. The second was several pages of his English translation of the *Continuo* related to Conclusions 38 to 40 sent to me by John Murdoch. Both shed some light on the argument given and I pause to express my thanks for these sendings.

Youschkevitch does not directly address the Conclusions in question, but in discussing Bradwardines's other works cites the *Geometria speculativa* (*Theoretical Geometry*). He mentions that, in reference to ibn Rushd (known as Averroës in the west), Bradwardine studied the problem we covered in section 5 of filling space with congruent regular polyhedra. There we read that ibn Rushd believed that such was possible not only with cubes, but also with tetrahedra. Although Youschkevitch does not report on what Bradwardine believed[44], this reinforced my belief that Bradwardine had something like Theorem 5.5 at the back of his mind in establishing Conclusion 38 and its corollary. However, it must be admitted that the dates of composition of the *Continuo* and *Geometria speculativa* are unknown. His interest in space-filling polyhedra may have stemmed from Conclusion 38, or it may even have stemmed from reading ibn Rushd and he never made the connexion at all.

As regards the *Continuo* itself, Youschkevitch offers another example of Bradwardine's reasoning. It is the usual "cardinality = measure" fallacy, this time applied to a semi-circle and a diameter. If continua consist of points, then the semicircle equals the diameter, for, dropping perpendiculars from the points of the semicircle to the diameter sets up a correspondence between the points on the semicircle and those on the diameter. Bradwardine rejects this conclusion both on theoretical grounds (in geometry one proves the semicircle to be greater than the diameter) and on empirical grounds (anyone who has ever measured it knows the semicircle to be greater than the diameter). Bradwardine was after a total rejection of indivisibles and sought

---

[44] I think the most annoying thing about secondary sources is not that they contain errors or off-the-wall interpretations, but that they NEVER include the specific information one is looking for.

to show them inconsistent not only with Euclidean geometry, but with all the known sciences.

Getting back to Conclusion 38, i.e. Lemma 4.2, Bradwardine makes no mention of regular polygons or polyhedra. He does draw a bunch of squares, as in *Figure 21*, but they represent points, not actual squares. His argument is two-fold. First, the point C has 8 immediate neighbours. For, starting with points in the 8 directions and drawing lines from them to C, each line segment being a continuum, we see that C has immediate neighbours in each of the 8 directions[45].

To prove that C has no other immediate neighbours, he starts with the line segment ACB with A, B immediate neighbours of C, and "superposes" HDG and IEF, with H and A, D and C, G and B, I and A, E and C, and F and B immediate neighbours. Any other immediate point would have to lie on one of the lines HDG, GBF, IEF, or HAI and be between two adjacent points on one of these lines. That this is impossible is Conclusion 5, which asserts a point to have only one immediate neighbour in a single direction on a straight line — a conclusion appealed to above in trying to rule out triangular atoms with reference to *Figures 23* and *24* on page 78.

"Superposition" is a term necessitated by philosophical subtleties that needn't concern us here. We can rephrase the argument more mathematically as follows. One starts with a point C and immediate neighbours A, B, D, E on perpendicular lines ACB and DCE. Through A and B draw lines parallel to DCE and through D and E lines parallel to ACB. Let F, G, H, I be the points of intersection of these lines as in *Figure 21*.

The existence of 8 immediate neighbours is established as before by considering the 8 directions given. A, B, D, and E are such by choice. F, G, H and I are also immediate neighbours as is seen by noting that any point immediate to C lies on one of the lines HDE, ACB, IEF, and one of HAI, BCE, GBF. The points of intersection, other than C, are A, B, D, E, F, G, H, I (whence there are at most 8).

The proof that a point X immediate to C lies on one of HDE, ACB, IEF runs as follows. Drop a perpendicular from X to DCE and look at the point Y of intersection with DCE. Y cannot come between C and one of its immediate neighbours D and E, whence Y must equal one of D, C, and E. Thus the line XY must coincide with HDE, ACB or IEF as Y is D, C, or E, i.e. X lies on one of HDE, ACB and IEF.

Similarly, X lies on one of HAI, BCE, and GBF.

As for the proof of Conclusion 40, i.e. Theorem 4.4, let me quote Murdoch's translation. (The bracketed expressions are Murdoch's explanatory interpolations.)

Let AB be a straight line cut perpendicularly at its midpoint C by [the straight line] DE. Then, by [Conclusion] 38, there are only 8

---

[45] Thus my speculation on page 78 that "only 8" should perhaps read "at most 8" was incorrect. I had translated correctly.

points immediately surrounding C, and 4 of these are in 4 lines, viz., in AC, DC, BC, EC. Therefore, since there are 4 [points] left over, they exist in the 4 angles, and consequently there is only one [point] in any one of the angles. Moreover, there cannot be an angle smaller than [an angle] of one point and, by Conclusion 1, one point cannot be smaller than another. Thus the first part...

This is a little clearer than my attempted translation, particularly in identifying the "earlier Conclusion" as Conclusion 1, but it doesn't clarify the mathematics behind the argument. He certainly doesn't seem to be suggesting the non-existence of the points F, G, H, I as I surmised earlier.

To make sense out of this, we must look to Conclusion 39, which I had not translated. It is even stranger than Conclusion 38: Every circle has exactly 8 points, every straight line exactly 3. I can illustrate the argument by showing that the point K in *Figure 21* does not exist. If K existed, there would be a point on the line CK adjacent to C. But the only points adjacent to C are A, B, D, E, F, G, H, I and none of them is on the line CK.

The same reasoning shows in fact that there are no points in the plane other than C and its 8 neighbours. Thus, e.g., G is the only point inside $\angle DCB$. Thus all the assertions of Bradwardine's proof make some sort of sense. What is not clear is why $\angle DCG$ and $\angle GCB$ do not exist. The only thing that comes to mind is that there are no points trapped inside of them and they might therefore be deemed as not constituting true angles.

Whatever decision one makes about the meaning and validity of Conclusion 40, i.e. Theorem 4.4, it is clear that Bradwardine has established the inconsistency of Harclay's adjacent points with the Euclidean axioms. That this can be seen immediately by citing the Euclidean construction of the midpoint between any two points ought not to mislead us into dismissing Bradwardine's results as pointless[46]. Rather, if one likes the arguments—and I do—, one should try to attach some significance to them by finding some context in which his results have a positive meaning and his arguments establish something other than contradictions. Our discussion of tilings of the plane provides an heuristic interpretation of the result, but does not model his proofs because the additional conclusion that there are only 9 points does not hold.

---

[46] Here I cannot resist quoting Hermann Hankel's judgment of Bradwardine and his contemporaries: "The presentation of these thoughts is however often so poor and what little is achieved is concealed in such a disorderly mess of scholastic subtleties and mathematical trivialities, that we mostly come to feel even a blind chicken can occasionally find a kernal of corn". (*Zur Geschichte der Mathematik im Alterthum und Mittelalter*, B.G. Teubner, Leipzig, 1874, pp. 351 - 352.)

4

The Construction Problems of Antiquity

1 Some Background

The classical construction problems of antiquity— squaring the circle, duplicating the cube, and trisecting the angle— cannot be solved by ruler and
compass construction alone. As shown already by the Greeks, these constructions can be effected if one augments one's tool set. Hippias invented a curve
called the quadratrix from a drawing of which the circle can be squared by
further application of ruler and compass. Drawing the quadratrix, on the other
hand, is another prolem. One can, however, duplicate the cube and trisect the
angle using curves that can be drawn by mechanical devices. From the Greeks
we have the conchoid of Nicomedes and the conic sections of Menæchmus;
among the Arabs, al-Khayyami used conic sections to solve related algebraic
problems; and in modern times we can point to Descartes who also relied on
conics. The simplest device for angle trisection is a ruler on which one is allowed to make two marks to measure a fixed distance with.[1] Today's history
texts not only cite these results, but include all the details and there is no
need to discuss them here. What the textbooks do not include and merits
inclusion here is the proof of impossibility.

There is not one proof, but two. The impossibilities of duplicating the cube
or trisecting the angle translate into algebraic questions amenable to algebraic
treatment; the impossibility of squaring the circle translates into an algebraic
problem best handled by analytic techniques. The first proof is due to Pierre
Wantzel[2] (1837), and the latter to Ferdinand Lindemann (1882). Wantzel's
results have been swallowed up by Galois Theory and are usually presented

---

[1] The rules for ruler and compass construction do not allow for marked rulers.
Thus, some prefer the term "straightedge" to "ruler".

[2] Pierre Laurent Wantzel is one of those rareties in mathematics, the solver of a
famous problem and an obscurity at the same time. His result is one of the most
famous in all of mathematics and yet he did not rate an entry in the *Dictionary
of Scientific Biography*.

to the student in a graduate course on abstract algebra; Lindemann's proof is generally reserved for advanced courses. This depth probably explains why the history books do not include proofs of these results. Wantzel's results, however, can be given quite elementary proofs that really do not use much algebra at all and I cannot imagine not proving them in class. Hence I include proofs of Wantzel's results here.

Before proceeding, let me expand a bit on the history of Wantzel's results. This story begins with Carl Friedrich Gauss, who made an incredible discovery in his youth: the regular 17-gon can be constructed by ruler and compass. This was the first new construction of a regular polygon since Euclid. When Gauss published this result in his famous *Disquisitiones Arithmeticae (Arithmetical Investigations)* in 1801, he actually proved that the regular p-gon can be constructed for any prime p of the form $2^{2^n} + 1$ [3] and warned the reader not to attempt such a construction for primes not of this form. He stated, but gave no hint of a proof, that the construction could not be given for other primes. When Wantzel published his paper[4] in 1837, he placed the nonconstructibility of the regular 7-gon alongside the impossibility of the duplication of the cube and the trisection of the angle.

Wantzel's paper does not seem to have been all that well-known. When the Danish mathematician Julius Petersen published the proof in a textbook in 1877[5], he made no mention of Wantzel[6]. In 1895 the American mathematician James Pierpont, apparently having rediscovered the proof independently, published it in a paper significantly titled "On an undemonstrated theorem of the Disquisitiones Arithmeticae"[7]. Pierpont makes no mention of Wantzel, the duplication of the cube, or the trisection of an angle. As he cites explicitly the constructibility of the regular 3-gon and the nonconstructibility of the regular 9-gon, from which facts the nontrisectability of the 120 degree angle follows immediately, this would seem a strange oversight. However, that very same year, Felix Klein published his famous lectures on the subject[8] and therein he states, regarding the duplication of the cube and the nontrisectability of the angle, that their impossibility is "implicitly involved in the Galois theory as presented today in treatises on higher algebra". This suggests that the proof of impossibility by appeal to Galois theory was well-known— and Pierpont

---

[3] Some authors write $2^n + 1$. Since a number of this form is prime only when n is a power of 2, the two different-looking formulations are equivalent.

[4] "Recherches sur les moyens de reconnaître si un Problème de Géometry peut se résoudre avec la règle et le compas", *Journal de Mathématiques* 2 (1837), pp. 366 - 372.

[5] Cf. footnote 15 below. Section 3 on conic sections discusses Petersen's work in more detail.

[6] However, he did mention Wantzel in his dissertation. In a discussion list on the history of mathematics on the Internet, Robin Hartshorne noted that there is a gap in Wantzel's proof that Petersen was the first to fill.

[7] *Bulletin of the Ametican Mathematical Society* 2 (1895), pp. 77 - 83.

[8] *Famous Problems of Elementary Geometry*, cited in the Bibliography.

knew that subject well. So we can probably assume his failure to mention these other algebraic impossibility results not to be an oversight but his unwillingness to prove known results in a research paper. However, the proofs given by Wantzel, Petersen, Pierpont, and Klein are all far more elementary than proof by appeal to Galois theory, and were not very widely known: Klein, for example, continues his reference to the fact that impossibility results are implicit in Galois theory with the observation that, "On the other hand, we find no explicit demonstration in elementary form unless it be in Petersen's textbooks, works which are also noteworthy in other respects".

The reader who sticks with this chapter to the bitter end will get the full flavour of the proof by Wantzel *et alia* in section 5, below. I present a much simplified proof in the immediately following section[9] [10]. Section 3 then presents some interesting results of Petersen and Laugwitz dealing with conic sections. This is followed by section 4, where I might be deemed to go too far. However, I wanted to give a sense of how far one can go with the most elementary means and where higher theory begins to kick in. Sections 3 and 4 may be read independently of each other immediately on completing section 2. Finally, section 6 proves one of Petersen's more general results and section 7 finishes our discussion with a few comments.

2 Unsolvability by Ruler and Compass

The classical construction problems of antiquity— squaring the circle, duplicating the cube, and trisecting the angle— cannot be solved by ruler and compass construction alone. The proof of this is algebraic in nature and in the case of the latter two problems is not too difficult. Like any good proof, it begins with a few definitions:

2.1 Definition. *A subset $F \subseteq \mathbb{R}$, the set of real numbers, is called a* number field *if it satisfies the following:*

---

[9] My exposition in the following section follows three papers of Detlef Laugwitz: "Unlösbarkeit geometrischer Konstruktionsaufgaben— Braucht man dazu moderne Algebra?", in: B. Fuchssteiner, U. Kulisch, D. Laugwitz, and R. Liedle, eds., *Jahrbuch Überblicke Mathematik, 1976,* Bibliographisches Institut, Mannheim, 1976.
"Eine elementare Methode für Unmöglichkeitsbeweise bei Konstruktionen mit Zirkel und Lineal", *Elemente der Mathematik* 17 (1962), pp. 54 - 58.
"Eine mit Zirkel und Lineal nicht lösbare Kegelschnittaufgabe", *Elemente der Mathematik* 26 (1971), pp. 135 - 136.

[10] Two English language sources for the basic proof are Richard Courant and Herbert Robbins, *What is Mathematics?,* Oxford University Press, Oxford, 1941, and Charles Robert Hadlock, *Field Theory and Its Classical Problems,* Mathematical Association of America, 1978. Hadlock cites Edmund Landau on Ludwig Bieberbach's authority as the source for this proof. He also presents a variant of Lindemann's proof of the transcendence of π from which it follows that the circle cannot be squared.

i. $0, 1 \in F$

ii. if $x, y \in F$ then $x + y, xy, x - y \in F$

iii. if $x, y \in F$ and $y \neq 0$ then $x/y \in F$.

Notice that, because F is assumed closed under multiplication, condition (iii) is equivalent to the following:

$$\text{if } y \in F \text{ and } y \neq 0 \text{ then } 1/y \in F.$$

For,

$$x/y = x \times 1/y.$$

2.2 Example. The set $\mathbb{Q}$ of rational numbers is a number field.

To see this, recall that a real number is rational if it can be written in the form m/n where m is an arbitrary integer and n is a positive integer. Then observe:

$$0 = 0/1$$

$$1 = 1/1$$

$$a/b + c/d = \frac{ad + cb}{bd}$$

$$a/b - c/d = \frac{ad - cb}{bd}$$

$$a/b \times c/d = \frac{ac}{bd}$$

$$\frac{a/b}{c/d} = a/b \times d/c = \frac{ad}{bc}.$$

In each case the result is a number of the form m/n with m integral and n positive integral.

2.3 Example. Let F be a number field and suppose $\alpha \in F$ is a positive real number such that $\sqrt{\alpha} \notin F$. Let $F[\sqrt{\alpha}]$ denote the set of all real numbers of the form $a + b\sqrt{\alpha}$, where $a, b \in F$. Then $F[\sqrt{\alpha}]$ is a number field.

As with the previous example, observe that:

$$0 = 0 + 0\sqrt{\alpha}$$

$$1 = 1 + 0\sqrt{\alpha}$$

$$(a + b\sqrt{\alpha}) + (c + d\sqrt{\alpha}) = (a + c) + (b + d)\sqrt{\alpha}$$

$$(a + b\sqrt{\alpha}) - (c + d\sqrt{\alpha}) = (a - c) + (b - d)\sqrt{\alpha}$$

$$(a + b\sqrt{\alpha}) \times (c + d\sqrt{\alpha}) = (ac + bd\alpha) + (ad + bc)\sqrt{\alpha}$$

$$\frac{1}{a + b\sqrt{\alpha}} = \frac{a - b\sqrt{\alpha}}{a^2 - b^2\alpha} = \frac{a}{a^2 - b^2\alpha} + \frac{-b}{a^2 - b^2\alpha}\sqrt{\alpha},$$

where we have used the usual process of rationalising the denominator in the last step. Again, the closure of F under the arithmetic operations tells us that the coefficients in the expressions on the right hand sides of these equations belong to F. [The one subtlety is that the denominator in the right side of the last equation is not 0: if it were, we would have

$$\alpha = a^2/b^2$$

whence $\sqrt{\alpha}$ would be a/b or $-a/b$, an element of F, contrary to assumption.]

Note that we can drop the assumption that $\sqrt{\alpha} \notin F$ in defining $F[\sqrt{\alpha}]$ and still obtain a field. However, if $\sqrt{\alpha} \in F$, the field obtained is not new, but coincides with F.

2.4 Definition. *A real number β is* constructible *if we can find a chain of number fields $\mathbb{Q} = F_0 \subseteq F_1 \subseteq \ldots \subseteq F_n$, where $\beta \in F_n$ and each F_{i+1} is of the form $F_i[\sqrt{\alpha_i}]$ for some $\alpha_i \in F_i$.*

More intuitively, β is constructible if it can be constructed in finitely many steps from rational numbers by the usual arithmetic operations and the extraction of square roots of positive numbers. For example, $2 + \sqrt{1 + \sqrt{3}}$ is constructible. We can easily exhibit a chain of fields for it:

$$\mathbb{Q} \subseteq \mathbb{Q}[\sqrt{3}] \subseteq (\mathbb{Q}[\sqrt{3}])[\sqrt{1 + \sqrt{3}}].$$

Such a chain need not be unique. For example, for $\sqrt{8} + \sqrt{15}$ we have the two chains:

$$\mathbb{Q} \subseteq \mathbb{Q}[\sqrt{8}] \subseteq (\mathbb{Q}[\sqrt{8}])[\sqrt{15}]$$

$$\mathbb{Q} \subseteq \mathbb{Q}[\sqrt{15}] \subseteq (\mathbb{Q}[\sqrt{15}])[\sqrt{8}].$$

A bit less obvious is the chain for $\sqrt{8} + \sqrt{18}$:

$$\mathbb{Q} \subseteq \mathbb{Q}[\sqrt{50}].$$

This last one works because $\sqrt{8} + \sqrt{18} = \sqrt{50}$. Moreover, because $50 = 2 \times 5^2$, this last chain is the same as

$$\mathbb{Q} \subseteq \mathbb{Q}[\sqrt{2}].$$

Those familiar with set theory will recognise that there are only countably many constructible numbers and uncountably many real numbers, whence most real numbers are not constructible. Concrete examples can be given. $\sqrt[3]{2}$ is not constructible. However, as the example of $\sqrt{8} + \sqrt{18}$ shows, the fact that a number is not written in a simpler form is no guarantee that it cannot be rewritten in a simpler form. Proving $\sqrt[3]{2}$ not to be constructible requires some thought.

The relevance of constructible real numbers to the construction problems of antiquity is that the constructible real numbers can be used to construct a model of Euclidean geometry. In describing this I shall be brief as I think it is fairly intuitive and the details are not too difficult to fill in. The fun stuff is with the algebra.

The basic idea is to restrict the usual coordinate plane to those points whose coordinates are both constructible. We may call such points *constructible*. A *constructible straight line* is the set of constructible points on a straight line in the usual plane determined by two constructible points. And a *constructible circle* is a circle whose centre is a constructible point and whose radius is a constructible number. Verifying that the constructible points, lines and circles satisfy the Euclidean axioms is a bit tricky because Euclid hides many of his assumptions in the Definitions and these Definitions are spread throughout the 13 books, but basically the verification boils down to the fact that finding the intersections of two constructible lines, or a constructible line and a constructible circle, or two constructible circles reduces to solving either a linear equation or a quadratic equation with constructible coefficients. Solving a linear equation over a given number field requires only closure under the arithmetic operations and can be done within that field; solving the quadratic may require the adjunction of a square root— whence the necessity of passage from a field F to a field $F[\sqrt{\alpha}]$.

The ruler and compass constructions come in as follows: laying out a straight line given by two constructible points gives rise to a linear equation, and drawing a circle of a given constructible radius centered at a given constructible point amounts to formulating a quadratic equation. Thus the lines, circles, and points constructed by ruler and compass are all constructible lines, circles, and points.

To show that a given object cannot be constructed, one need only show that the object allows the "construction" of a non-constructible number. For example in doubling the cube of a constructible line segment one shows that the length of the doubling segment is not constructible. It suffices to show this for a single example, so we may assume a cube of edge 1 and show that the length $\sqrt[3]{2}$ of the doubling segment is not a constructible number. To show that a given angle cannot be constructed, it suffices to show its sine, cosine or tangent is not a constructible number.

The tasks of exhibiting the nonconstructible natures of $\sqrt[3]{2}$ and the trisecting angles of certain angles are rendered fairly easy because these numbers satisfy certain cubic equations and cubic equations have a special property *vis-à-vis* constructibility.

2.5 Theorem. *Consider a polynomial of the form $P(X) = X^3 + aX^2 + bX + c$, with $a, b, c \in \mathbb{Q}$. If P has a constructible solution, then P has a rational solution as well.*

Proof. If P has a rational solution we are finished. Suppose then that P has no rational solution. We shall derive a contradiction from that assumption.

Let β be a constructible root of P: $P(\beta) = 0$. Then there is a chain of number fields $\mathbb{Q} = F_0 \subseteq F_1 \subseteq \ldots \subseteq F_n$, where each F_{i+1} extends F_i by the adjunction of the square root of an element of F_i and where $\beta \in F_n$. Assume n is minimal with this property, i.e. that we have exhibited the shortest chain by which such a constructible root of P can be found. Write $\beta = p + q\sqrt{\alpha}$, where $\alpha \in F_{n-1}$ but $\sqrt{\alpha} \notin F_{n-1}$.

$$
\begin{aligned}
P(\beta) =& p^3 + 3p^2 q\sqrt{\alpha} + 3pq^2\alpha + q^3\alpha\sqrt{\alpha} \\
&+ a(p^2 + 2pq\sqrt{\alpha} + q^2\alpha) + b(p + q\sqrt{\alpha}) + c \\
=& (p^3 + 3pq^2\alpha + ap^2 + aq^2\alpha + bp + c) \\
&+ (3p^2 q + q^3\alpha + 2apq + bq)\sqrt{\alpha}.
\end{aligned}
$$

Thus we have $P(\beta) = A + B\sqrt{\alpha}$ for some $A, B \in F_{n-1}$. Now both A, B must be 0. For, if $B \neq 0$, then $\sqrt{\alpha} = -A/B \in F_{n-1}$, contrary to assumption. Thus $B = 0$ and since $P(\beta) = 0$, it follows that $A = P(\beta) = 0$.

But a similar calculation shows

$$
\begin{aligned}
P(p - q\sqrt{\alpha}) =& (p^3 + 3pq^2\alpha + ap^2 + aq^2\alpha + bp + c) \\
&- (3p^2 q + q^3\alpha + 2apq + bq)\sqrt{\alpha} \\
=& A - B\sqrt{\alpha} = 0.
\end{aligned}
$$

For convenience we write $\overline{\beta}$ for $p - q\sqrt{\alpha}$.

We are almost done. Letting γ be the third real root of P, and factoring we have:

$$
\begin{aligned}
P(X) =& (X - \beta)(X - \overline{\beta})(X - \gamma) \\
=& X^3 + (-\beta - \overline{\beta} - \gamma)X^2 + (\beta\overline{\beta} + \beta\gamma + \overline{\beta}\gamma)X - \beta\overline{\beta}\gamma.
\end{aligned}
$$

From this we see that $a = -\beta - \overline{\beta} - \gamma$, i.e.,

$$
\gamma = -(a + \beta + \overline{\beta}) = -(a + p + q\sqrt{\alpha} + p - q\sqrt{\alpha}) = -(a + 2p),
$$

which is an element of F_{n-1} contrary to the minimality assumption on n. Therefore the assumption that P had no rational root was incorrect and we conclude that P has such a root.(If you don't like the argument by contradiction or the appeal to minimality, we have shown that if F_n possesses a constructible root β of P, then F_{n-1} possesses one. The same argument then shows F_{n-2} has such a root. Repeating the process eventually puts one into $F_0 = \mathbb{Q}$.) $\qquad\square$

2.6 Corollary. *The cube cannot be duplicated by ruler and compass construction alone.*

Proof. Doubling the cube on a side of length 1 means constructing a line segment of length $\sqrt[3]{2}$. This requires $\sqrt[3]{2}$ to be constructible. But $\sqrt[3]{2}$ satisfies the equation $X^3 - 2 = 0$ and, if $\sqrt[3]{2}$ were constructible, this would have a rational solution, which it clearly does not. $\qquad\square$

2.7 Corollary. *The angle cannot be trisected by ruler and compass construction alone.*

Proof. This is a bit trickier and requires some trigonometry, specifically the Addition Formulæ for sines or cosines. For any angle θ, we have

$$\cos 3\theta = \cos(2\theta + \theta) = \cos 2\theta \cos \theta - \sin 2\theta \sin \theta$$
$$= (\cos^2 \theta - \sin^2 \theta) \cos \theta - 2 \sin^2 \theta \cos \theta$$
$$= \cos^3 \theta - (1 - \cos^2 \theta) \cos \theta - 2(1 - \cos^2 \theta) \cos \theta$$
$$= \cos^3 \theta - \cos \theta + \cos^3 \theta - 2 \cos \theta + 2 \cos^3 \theta$$
$$= 4 \cos^3 \theta - 3 \cos \theta.$$

For $3\theta = 60$ degrees, i.e. for θ equal to 20 degrees, we have

$$\frac{1}{2} = 4 \cos^3 \theta - 3 \cos \theta,$$

i.e. $\cos \theta$ is a solution to the equation

$$\frac{1}{2} = 4X^3 - 3X,$$

whence to

$$X^3 - \frac{3}{4}X - \frac{1}{8} = 0. \tag{1}$$

Now if we could trisect the 60 degree angle by ruler and compass, then $\cos 20$ would be a constructible root of this last equation and by the Theorem it would have a rational root.

One can use ordinary college algebra to verify that (1) has no rational solution. Multiplying by 8, the equation becomes

$$8X^3 - 6X - 1 = 0.$$

In college algebra we learn that any rational solution to this equation must have its numerator a divisor of -1 and its denominator a positive divisor of 8. This yields $\pm 1, \pm 1/2, \pm 1/4, \pm 1/8$ as candidates. Testing them one at a time reveals none of them to be roots.

This finishes the proof, but before I close it with the proof-closing box, I want to take another look at this last step. The number of candidates to test for rational roots is not that great that one cannot do the calculations via the calculator or even by hand, but it is great enough to make one think about short cuts. We will consider such in the section on computations in the chapter on the cubic equation. For now, we don't need to go there because the simple substitution $2X = Y$ will transform the equation into

$$Y^3 - 3Y - 1 = 0,$$

which has only ± 1 as candidates for a rational solution and neither of these satisfies the equation.   □

Note that, had we started with sines instead of cosines and attempted to trisect a 30 degree angle, we would have gotten a similar equation showing that $\sin 10$ is not a constructible number. Of course, we don't need to do this because it follows from the nonconstructibility of the 20 degree angle that the 10 degree angle is also nonconstructible— for otherwise we could construct the 10 degree angle and double it. Also, because we can bisect angles, the 40 degree angle cannot be constructed by ruler and compass.

2.8 Corollary. *The regular 7-sided polygon cannot be constructed by ruler and compass.*

Proof. Oddly enough we can get a cubic equation out of this problem. If one inscribes the regular 7-sided polygon inside the unit circle in such a way that one of the vertices falls on the point $(1,0)$, the other six vertices will fall on points $(\cos(360k/7), \sin(360k/7))$ where the integer k runs from 1 through 6. In terms of complex numbers and radians, these are the points $\cos 2\pi k/7 + i \sin 2\pi k/7$, all of which are 7th roots of 1, i.e. are solutions to the equation:
$$X^7 - 1 = 0.$$
Eliminating the trivial root $X = 1$ by dividing by $X - 1$, they are solutions to the equation

$$X^6 + X^5 + X^4 + X^3 + X^2 + X^1 + 1 = 0. \tag{2}$$

If we divide by X^3, we get

$$X^3 + X^2 + X + 1 + 1/X + 1/X^2 + 1/X^3 = 0,$$

and then

$$(X^3 + 1/X^3) + (X^2 + 1/X^2) + (X + 1/X) + 1 = 0.$$

If we now set $Y = X + 1/X$, we see that

$$X^3 + 1/X^3 = Y^3 - 3Y$$
$$X^2 + 1/X^2 = Y^2 - 2$$
$$X + 1/X = Y.$$

Thus (2) becomes
$$Y^3 + Y^2 - 2Y - 1 = 0, \tag{3}$$

a nice cubic equation to which to apply the Theorem. From college algebra we know that the only rational candidates for solutions to this equation are ± 1, which quick computations quickly rule out.

Now what is a solution to (3)? It is a number of the form

$$(\cos 2\pi k/7 + i\sin 2\pi k/7) + (\cos 2\pi k/7 - i\sin 2\pi k/7) = 2\cos 2\pi k/7.$$

As this is not a constructible number, neither is $\cos 2\pi k/7$ and, in particular, $\cos 2\pi/7$ is not constructible. But this is the x-coordinate of the first vertex of the regular 7-gon of side 1. Hence we cannot construct this 7-gon by ruler and compass. □

Corollary 2.8 is no longer one of the famous trio of construction problems of antiquity, even though its antiquity can readily be vouchsafed. When possible, Euclid provided constructions for regular polygons and even gave the construction for inscribing a regular 15-gon inside a circle. Needless to say, he did not construct the regular 7-gon. Nor could he construct the regular 9-gon, as I leave to the reader to prove.

There are plenty of additional construction problems one can pose. In 1947, P. Buchner[11] proved the impossibility of the following problem put to him by an engineer:

2.9 Problem. *Given two lengths, a, b and a radius r, construct a triangle with two sides of lengths a and b and an inscribed circle of radius r.*

Laugwitz gave a counterexample to a special case of the problem:

2.10 Corollary. *There are constructible values of $a, r > 0$ for which there is an isosceles triangle with two sides of length a and an inscribed circle of radius r, but for which no such triangle can be constructed using only ruler and compass.*

Proof. Consider the triangle ABC of *Figure 1*. Let α denote the angle at A, and c the remaining side. The line drawn from vertex A to the centre of the inscribed circle bisects α. Notice that

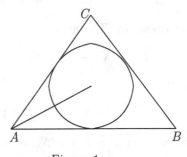

Figure 1

$$\tan\frac{\alpha}{2} = \frac{r}{c/2} = \frac{2r}{c}$$

$$\cos\alpha = \frac{c/2}{a} = \frac{c}{2a}.$$

We can express the cosine of an angle in terms of the tangent of its half-angle:

$$\frac{c}{2a} = \cos\alpha = \frac{1 - \tan^2(\alpha/2)}{1 + \tan^2(\alpha/2)} = \frac{1 - \left(\frac{2r}{c}\right)^2}{1 + \left(\frac{2r}{c}\right)^2} = \frac{c^2 - 4r^2}{c^2 + 4r^2}.$$

Cross-multiplying and simplifying, we have

$$c^3 - 2ac^2 + 4r^2c + 8ar^2 = 0,$$

---

[11] "Eine Aufgabe, die mit Zirkel und Lineal nicht lösbar ist", *Elemente der Mathematik 2* (1947), pp. 14 - 16.

which is cubic in c. If we can construct the figure in question then c is certainly a constructible number. Hence it suffices to choose a, r so that

i. the equation,

$$X^3 - 2aX^2 + 4r^2X + 8ar^2 = 0$$

has no constructible solution; and

ii. the triangle exists.

The second condition is met so long as a is sufficiently larger than r. Note, for example, that the choice of $a = 1, r = 1/2$ results in the equation

$$X^3 - 2X^2 + X + 2 = 0,$$

which has no rational, hence no constructible solution. However, in this case, a is too small for the triangle to inscribe a circle of radius $1/2$. [One can prove this rigorously by noting that if the triangle exists, it must satisfy

$$a^2 = \left(\frac{c}{2}\right)^2 + h^2, \ h \text{ being the height}$$

$$> \frac{c^2}{4} + (2r)^2$$

$$4a^2 > c^2 + 16r^2$$

$$4 > c^2 + 4, \text{ for } a = 1, r = 1/2$$

$$0 > c^2,$$

which is impossible.] On the other hand, for $a = 2, r = 1/2$, we get the equation

$$X^3 - 4X^2 + X + 4 = 0,$$

which also has no constructible solution, but for which the triangle exists.

2.11 Lemma. *Let $a \geq 3.61r$. Then a triangle with two sides of length a and an inscribed circle of radius r exists.*

Proof. Consider *Figure 2*, in which the grid lines are r units apart. Line FH has length $\sqrt{2^2 + 3^2}\, r = \sqrt{13}r = 3.605551205r$. It or any line of length $a >$ it or $> 3.61r$ can be used to form the left half of the desired figure by sliding the bottom end point F toward G while sliding H upwards until the line segment touches the circle.   □

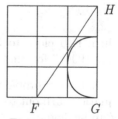

Figure 2

Since $a = 4r$ for $a = 2$ and $r = 1/2$, this completes the proof of the Corollary.   □

2.12 Remark. Laugwitz gives $a \geq 2\sqrt{3}r \approx 3.464r$ as a bound for the Lemma. The adequacy of this bound is readily established by first referring back to

Figure 1 and assuming ABC to be an equilateral triangle. For such some mild calculation reveals the ratio a/r to be exactly $2\sqrt{3}$, whence a triangle can be constructed for a/r in this or, by the argument of the Lemma, any higher ratio.

2.13 Remark. One can use Calculus to minimise the ratio. To do this, let β denote the half angle at C, x the length CD, and $b = a - x$. Then

$$\tan \beta = \frac{r}{x},$$

whence

$$x = \frac{r}{\tan \beta} = r\frac{\cos \beta}{\sin \beta}, \tag{4}$$

and

$$\sin \beta = \frac{b}{x + b}.$$

Using (4) and solving for b yields

$$b = \frac{r \cos \beta}{1 - \sin \beta},$$

whence

$$a = x + b = r \cos \beta \left[\frac{1}{\sin \beta} + \frac{1}{1 - \sin \beta} \right]$$

$$= r \cos \beta \left[\frac{1 - \sin \beta + \sin \beta}{\sin \beta(1 - \sin \beta)} \right] = \frac{r \cos \beta}{\sin \beta(1 - \sin \beta)}.$$

This gives

$$\frac{a}{r} = f(\beta) = \frac{\cos \beta}{\sin \beta - \sin^2 \beta},$$

and, after much simplification,

$$f'(\beta) = \frac{-\sin^3 \beta + 2\sin \beta - 1}{(\sin \beta - \sin^2 \beta)^2},$$

which is 0 close to $38°$ where a minimum value for a/r of approximately 3.330190677 for angles between 0 and 90 degrees is attained.

Not every problem that arises corresponds to a cubic equation. Should the equation turn out to be of the fourth degree, however, Theorem 2.5 has a sort of an analogue.

2.14 Theorem. *Let $P(X) = X^4 + aX^2 + bX + c$ be a reduced polynomial of degree 4 with a, b, c rational, and suppose P has no rational root. If P has a constructible root, then so does its cubic resolvent,*

$$Q(X) = X^3 + 2aX^2 + (a^2 - 4c)X - b^2.$$

Some words of explanation: A *reduced* n-th degree polynomial is one with no term of degree $n - 1$. Every polynomial

$$R(X) = a_0 X^n + a_1 X^{n-1} + \ldots + a_n$$

can be transformed into a reduced polynomial via the substitution

$$X = Y - \frac{a_1}{n a_0}.$$

(See the chapter below on the cubic equation.) The cubic resolvent is obtained as follows: Let $\alpha_1, \alpha_2, \alpha_3, \alpha_4$ be the roots of P and define[12]

$$\theta_1 = (\alpha_1 + \alpha_2)(\alpha_3 + \alpha_4)$$
$$\theta_2 = (\alpha_1 + \alpha_3)(\alpha_2 + \alpha_4)$$
$$\theta_3 = (\alpha_1 + \alpha_4)(\alpha_2 + \alpha_3).$$

Then

$$Q(X) = (X - \theta_1)(X - \theta_2)(X - \theta_3).$$

That this agrees with the polynomial Q of the statement of the Theorem is a matter of grubby computation, which I leave to the reader, noting only that

$$P(X) = X^4 - \left(\sum_i \alpha_i\right) X^3 + \left(\sum_{i<j} \alpha_i \alpha_j\right) X^2 - \left(\sum_{i<j<k} \alpha_i \alpha_j \alpha_k\right) X + \alpha_1 \alpha_2 \alpha_3 \alpha_4$$

$$(5)$$

$$Q(X) = X^3 - \left(\sum_i \theta_i\right) X^2 + \left(\sum_{i<j} \theta_i \theta_j\right) X - \theta_1 \theta_2 \theta_3.$$

Proof of the Theorem. Suppose α is a constructible root of P. Since α is not rational, α occurs first in some $F[\sqrt{\varepsilon}]$ of a chain of extensions of $\mathbb{Q}$: $\alpha = p + q\sqrt{\varepsilon}$, where $p, q \in F$, $\sqrt{\varepsilon} \notin F$ and (since $\alpha \notin F$) $q \neq 0$. As in the proof of Theorem 2.5, if we consider $\overline{\alpha} = p - q\sqrt{\varepsilon}$, we will see that $\overline{\alpha}$ is also a root of P.

Without loss of generality, we may assume $\alpha = \alpha_1$, $\overline{\alpha} = \alpha_2$. Now $\alpha_1 + \alpha_2$ is constructible, whence so is

$$\alpha_3 + \alpha_4 = -(\alpha_1 + \alpha_2),$$

this last equation following by (5) and the assumption that P is reduced. But then $\theta_1 = (\alpha_1 + \alpha_2)(\alpha_3 + \alpha_4)$ is a constructible root of Q. $\qquad \square$

So to check that a reduced fourth degree polynomial has no constructible root, it suffices to show that neither it nor its cubic resolvent has a rational root. We will give a sample application of this in the next section.

---

[12] The motivation for the definition of Q comes from Galois Theory. Cf. e.g. B.L. van der Wærden's *Moderne Algebra*. If you have the Heidelberg Taschenbuch edition, the specific reference is to section 64.

3 Conic Sections

In his textbook on geometrical constructions[13] Julius Petersen makes some
very interesting general claims about ruler and compass constructions, which
I quote here:

1. *There are no other curves than the conic sections, of which one can determine the intersection with a straight line by straight edge and compasses.*

2. *There are no other curves than the conic sections to which tangents may
   be drawn from any point by straight edge and compasses.*

3. *If we are able by straight edge and compasses to determine the points, in
   which any line of a pencil of lines intersects a curve,* (which does not pass
   through the vertex of the pencil), *the order of this curve must be a power
   of 2 and there will be at least two lines in the pencil, of which the points
   of intersection with the curve coincide in pairs.*

4. *There are no other curves than the circle and the straight line, of which
   the points of intersection with any circle can be determined by straight
   edge and compasses.*

Petersen does not include the proofs, but refers to the published form of his
dissertation of 1871[14] and to his textbook on the theory of algebraic equations[15]. He does mention that assertion 4 follows from 1 because the circle
and straight lines are the only curves which form conic sections by inversion.
In his book on the theory of equations, he concludes 1 immediately after deriving 3[16] and states that 2 follows from 1 by duality. He also proves 3 in this
latter book, but his proof is rather sparse and I haven't mastered it yet. One
of the difficulties is that, as stated, the results are simply false and one has to
figure out what he means by "curve" from the properties of a curve used in
the proof.

Consulting the book on the theory of equations narrows the meaning quite
a bit. He is clearly referring to *algebraic curves*, i.e. curves defined by a polynomial equation,

---

[13] J. Petersen, *Methods and Theories for the Solution of Problems of Geometrical
    Construction*, reproduced in *String Figures and Other Monographs*, Chelsea Pub-
    lishing Company, New York, 1960. Petersen wrote a number of popular, but often
    difficult textbooks that were translated into several languages.

[14] "Om Ligninger, der løses ved Kvadratrod med Anvendelse paa Problemers
    Løsning ved Passer og Lineal", C. Ferslew and Co., Copenhagen, 1871.

[15] J. Petersen, *Theorie der algebraischen Gleichungen*, Andr. Fred. Höst & Sohn,
    Kopenhagen, 1878. This originally appeared in 1877 in Danish, but the German
    edition seems to be the most accessible. In any event, it is the one I found.

[16] This led me to believe that 1 is a corollary to 3. However, I don't see how it
    follows. I discuss these matters further in section 6, below, where I discuss a
    proof of assertion 3.

$$P(X, Y) = 0,$$

and not the more general parametrically defined curves familiar to us from the Calculus. Indeed, with respect to assertion 1, we can construct lots of non-algebraic curves for which we can use ruler and compass to determine their intersections with straight lines: take any piecewise linear curve. Using circular arcs and line segments, we can even form continuously differentiable such curves.

By the same reasoning, we can see that Petersen almost requires his curves to be *irreducible* over the reals. For, if $P_1(X, Y)$ and $P_2(X, Y)$ are polynomials defining curves $P_1(X, Y) = 0$ and $P_2(X, Y) = 0$, respectively, with the property, say, that we can determine their intersections with straight lines by means of ruler and compass, then we can clearly do the same with the curve $P_1(X, Y) \cdot P_2(X, Y) = 0$, which is the union of the original curves. Now, even if we assume the original curves to be conic sections, the product curve need not be a conic section, but the union of two such— say a circle and a parabola. Thus, for the first statement to be valid, if P factors into non-constant terms, each of them must have the same graph, i.e. P is a constant multiple of a power of an irreducible polynomial, say $P = cP_1(X, Y)^n$. One may as well restrict one's attention to P_1 and assume irreducibility at the outset.

Finally, one must make some constructibility assumptions. If, say, a parabola intersects a line at a constructible point A and we rigidly slide the parabola a nonconstructible distance along the line, then the point of intersection of the new parabola with the given line is no longer a constructible point. Or we could keep the parabola fixed and slide the straight line until it intersects the parabola in a nonconstructible point[17]; we run into the same problem. We must either relativise the notion of constructibility or limit our attention to constructible lines and curves. I choose the latter option. The question is: we know what a constructible line is; what is a constructible curve?

Once we have chosen a unit, geometry becomes algebra: constructible points are points with constructible coordinates; constructible lines are lines determined by pairs of constructible points; and circles are those with constructible centres and radii. Constructible lines and circles can also be characterised algebraically as those lines and circles definable by equations with constructible coefficients. Still other curves may be characterised constructibly even though they cannot be constructed. A parabola, for example, can be defined to be the locus of points equidistant from a given point called the *focus* and a given line called the *directrix*. If the focus and directrix are constructible it is reasonable to refer to the parabola as constructible, or , at least, as constructibly determined, and to raise the question of what can constructibly be done with such a parabola. Using a focus, directrix and an *eccentricity* not equal to 1, one can also define the notions of constructible ellipse and con-

---

[17] By general set theoretic consideration, most points on the parabola are non-constructible.

structible hyperbola. In general, a non-circular conic section can be considered constructible if it is determined by constructible focus, directrix, and eccentricity. And, as with the line and the circle, if a conic section is constructible, it is definable by an equation with constructible real coefficients. That this is so is not a trivial matter, but one of some calculation.

Suppose we are given a directrix $AX + BY + C = 0$, a focus (D, F), and an eccentricity $E > 0$. The distance from a point (X, Y) to the focus is

$$\sqrt{(X - D)^2 + (Y - F)^2}.$$

Otto Hesse's formula for the distance from a point to a line tells us that the distance from (X, Y) to the directrix is

$$\frac{|AX + BY + C|}{\sqrt{A^2 + B^2}}.$$

Thus, (X, Y) is on the conic just in case

$$\sqrt{(X - D)^2 + (Y - F)^2} = E\frac{|AX + BY + C|}{\sqrt{A^2 + B^2}},$$

or, squaring both sides, if

$$(X - D)^2 + (Y - F)^2 = E^2\frac{(AX + BY + C)^2}{A^2 + B^2}.$$

Multiplying this out and collecting like terms converts this to the form,

$$aX^2 + bXY + cY^2 + dX + eY + f = 0, \tag{6}$$

where a, b, c, d, e, f are also constructible numbers.

Conversely, if a non-circular conic section is defined by equation (6) with constructible coefficients, the usual algebraic determination of the focus, directrix and eccentricity results in constructible numbers.

We don't have a general geometric description of higher degree curves, but we do have the algebraic expressions to specify them.

3.1 Definitions. *An* algebraic curve, *defined as the graph of a polynomial equation,*

$$P(X, Y) = 0,$$

is constructibly determined *if the coefficients of P are constructible real numbers. The curve is* irreducible *if P is irreducible over $\mathbb{R}$, i.e. if P does not factor over $\mathbb{R}$ into polynomials of lower degree.*

The temptation to use the simpler adjective "constructible" is great, but we must ask ourselves just how constructible a constructibly determined algebraic curve is. The following theorem is my interpretation of Petersen's first claim.

3.2 Theorem. *The constructibly determined irreducible algebraic curves for which the intersections with all constructible lines can be determined by ruler and compass are precisely the conic sections.*

I will not attempt to prove this, but will only prove that conic sections do have the property and give an example of a simple cubic that does not have the property.

Proof that conic sections have the property. Let the conic be given by the equation (6) with constructible coefficients and suppose a constructible line $gX + hY + i = 0$ is given.

If $h = 0$, then $g \neq 0$ and the line is vertical with equation $X = -i/g$. Plugging this into (6), one gets

$$a\left(-\frac{i}{g}\right)^2 + b\left(-\frac{i}{g}\right)Y + cY^2 + d\left(-\frac{i}{g}\right) + eY + f = 0,$$

a quadratic equation, the real solutions of which are obtainable from $a, \ldots, i$ using the field operations and square root extraction, whence they are constructible.

If $h \neq 0$, one can solve the linear equation for Y: $Y = -(g/h)X - (i/h)$. Plugging this into (6), one gets

$$aX^2 + bX\left(-\frac{g}{h}X - \frac{i}{h}\right) + c\left(-\frac{g}{h}X - \frac{i}{h}\right)^2 + dX + e\left(-\frac{g}{h}X - \frac{i}{h}\right) + f = 0.$$

This is quadratic in X and its real solutions are hence constructible.   □

3.3 Example. The intersection of the cubic $Y = X^3$ and a constructible straight line need not be a constructible point. For, the intersection with the line $Y = 2$ is the point $(\sqrt[3]{2}, 2)$.

My interpretation of Petersen's second assertion is the following:

3.4 Theorem. *The constructibly determined irreducible algebraic curves to which the tangent lines may be drawn by ruler and compass from any given constructible point are precisely the conic sections.*

Again, we will have to satisfy ourselves here with the proof of the positive assertion and a single counterexample

Proof that conic sections have the property. Suppose the conic is defined by (6), and (α, β) is a given constructible point. Suppose further the non-vertical line $Y = MX + B$ passes through (α, β) and is tangent to the conic. [The case of a vertical tangent line $X = \alpha$ follows from Theorem 3.2.] Because (α, β) is on the tangent line, $B = \beta - M\alpha$, whence the equation of the tangent line may be written $Y = M(X - \alpha) + \beta$.

There may be several lines passing through (α, β) that are tangent to the conic, but each tangent meets the conic in only one point. This point of intersection for the tangent under consideration satisfies

$$0 = P(X, M(X - \alpha) + \beta)$$
$$= aX^2 + bX\Big(M(X - \alpha) + \beta\Big) + c\Big(M(X - \alpha) + \beta\Big)^2$$
$$\quad + dX + e\Big(M(X - \alpha) + \beta\Big) + f$$
$$= aX^2 + bMX^2 - \alpha bMX + b\beta X$$
$$\quad + c\Big(M^2(X - \alpha)^2 + 2M\beta(X - \alpha) + \beta^2\Big)$$
$$\quad + dX + e(MX - M\alpha + \beta) + f$$
$$= \Big(a + bM + cM^2\Big)X^2$$
$$\quad + \Big(-\alpha bM + b\beta - 2\alpha cM^2 + 2\beta cM + d + eM\Big)X$$
$$\quad + \Big(\alpha^2 cM^2 - 2\alpha\beta cM + \beta^2 c - \alpha eM + \beta e + f\Big).$$

This is quadratic in X and will have an unique solution just in case the discriminant,

$$\Big(-2\alpha cM^2 + (-\alpha b + 2\beta c + e)M + b\beta + d\Big)^2$$
$$\quad - 4\Big(cM^2 + bM + a\Big)\Big(\alpha^2 cM^2 + (-2\alpha\beta c - \alpha e)M + \beta^2 c + \beta e + f\Big),$$

is 0. Viewed as a polynomial in M, this is ostensibly of degree 4. However, if we collect terms, the coefficient of M^4 is

$$4\alpha^2 c^2 - 4c\alpha^2 c = 0.$$

The coefficient of M^3 is

$$-2\alpha c(-\alpha b + 2\beta c + e) \cdot 2 - 4c(-2\alpha\beta c - \alpha e) - 4b\alpha^2 c$$
$$= 4\alpha^2 bc - 8\alpha\beta c^2 - 4\alpha ce + 8\alpha\beta c^2 + 4\alpha ce - 4\alpha^2 bc$$
$$= 0.$$

Thus M satisfies a quadratic equation with constructible coefficients, whence M is constructible. It follows that $B = \beta - M\alpha$ is also constructible by Theorem 3.2.   □

Once again, in place of the full converse we settle for a counterexample.

3.5 Example. The tangent lines from the point $(2, 2)$ to the cubic $Y = X^3$ are not constructible.

Proof. Since $Y' = 3X^2$, the slope of the tangent line from a point (x, y) on the curve to a point (α, β) in the plane, if such a tangent exists, satisfies

$$\frac{y - \beta}{x - \alpha} = 3x^2.$$

Thus the point (x, y) of tangency satisfies the equation

$$Y - \beta = 3X^2(X - \alpha) = 3X^3 - 3\alpha X^2$$

and, since $Y = X^3$ at this point,

$$X^3 - \beta = 3X^3 - 3\alpha X^2,$$

i.e.

$$2X^3 - 3\alpha X^2 + \beta = 0. \tag{7}$$

But, for $(\alpha, \beta) = (2, 2)$, this is equivalent to

$$X^3 - 3X^2 + 1 = 0, \tag{8}$$

which has no rational, hence no constructible solution.

For the sake of completeness, it should be noted that equation (8) has three distinct real roots, x_1, x_2, x_3 and the three lines tangent to the curve $Y = X^3$ at the points $(x_1, x_1^3), (x_2, x_2^3), (x_3, x_3^3)$ all intersect at $(2, 2)$— as one may readily verify with any good graphing calculator.   □

One who finds the proliferation of tangent lines in this Example messy can choose $(\alpha, \beta) = (0, 1)$ so that (7) becomes

$$2X^3 + 1 = 0$$

with unique nonconstructible real solution $X = \sqrt[3]{-1/2}$, and hence only one tangent line extending from the curve to $(0, 1)$.

Backing up a bit, let us reconsider Theorem 3.2. According to it, the points of intersection of any constructible straight line with any constructibly detemined conic section were constructible. What happens if we replace the constructible straight lines by constructible circles? According to Petersen's fourth claim, the points of intersection need no longer be constructible.

3.6 Theorem. *The constructibly determined irreducible algebraic curves for which the intersections with all constructible circles can be determined by ruler and compass are precisely the constructible lines and circles.*

For this theorem, we already know the positive result. And again I only wish to illustrate the converse with a counterexample. In this case, however, the illustration does appear in the history texts. For, it consists of showing how to solve any cubic equation by means of conic sections à la Menæchmus, al-Khayyami, and Descartes.

3.7 Example. Any cubic equation can be solved by finding the points of intersection of a circle and a parabola.

Proof. Assume the cubic equation, after reduction[18], assumes the form

---

[18] Refer back to the end of the preceding section for the reduction of a cubic. We will have more to say on this matter in the chapter on the cubic equation.

$$X^3 + pX + q = 0. \tag{9}$$

We introduce the parabola

$$Y = X^2 + \frac{p-1}{2}$$

and the circle

$$\left(X + \frac{q}{2}\right)^2 + Y^2 = \left(\frac{p-1}{2}\right)^2 + \left(\frac{q}{2}\right)^2.$$

If $p = 1$ and $q = 0$, the "circle" degenerates to the point $(0,0)$. But in this case, the equation (9) is $X^3 + X = 0$ with only the solution $X = 0$ corresponding exactly to this point. Hence, we may ignore this case in what follows.

The points of intersection satisfy the substituted equation,

$$\left(X + \frac{q}{2}\right)^2 + \left(X^2 + \frac{p-1}{2}\right)^2 = \left(\frac{p-1}{2}\right)^2 + \left(\frac{q}{2}\right)^2.$$

Multiplying this out,

$$X^2 + qX + \left(\frac{q}{2}\right)^2 + X^4 + (p-1)X^2 + \left(\frac{p-1}{2}\right)^2 = \left(\frac{p-1}{2}\right)^2 + \left(\frac{q}{2}\right)^2,$$

i.e.

$$X^4 + pX^2 + qX = 0,$$

whence $X = 0$ or $X^3 + pX + q = 0$. □

3.8 Example. The points of intersection of the parabola $Y = X^2 - 1/2$ and the circle $(X - 1)^2 + Y^2 = 5/4$ are not all constructible.

Proof. These arise from Example 3.7 by choosing $p = 0, q = -2$ corresponding to the cubic equation $X^3 - 2 = 0$. The point of intersection other than $X = 0$ is given by $X = \sqrt[3]{2}$, which is not constructible. □

Petersen's very general results concerning constructibility and conics do not exhaust all the possibilities. Laugwitz provides the following additional result.

3.9 Theorem. *The distance from a constructible line to a constructibly determined conic section is constructible, but the distance from a constructible point to a constructibly determined conic need not be.*

Proof of the positive assertion. But for the necessity of sorting out the exceptional cases, the proof is simple and straightforward. Let the equation of the conic be given by (6) and let the straight line have equation $AX + BY + C = 0$. Rotating the plane through a constructible angle if necessary, we can assume the line is neither vertical nor horizontal, i.e. we can assume $A, B \neq 0$ and find the nonzero slope $-A/B$ of the line.

We can also assume the line not to intersect the conic section, else we could simply appeal to Theorem 3.2. We may also assume the conic is not degenerate, i.e. it is not a point, a line, or a pair of lines, as the distance is easily determined in these cases. Thus, we have a real conic and a straight line not intersecting it. The shortest distance between them will be given by a line segment perpendicular to both the conic and the straight line. The tangent to the conic at a[19] point on the conic closest to the straight line is parallel to said line and will have the slope $-A/B$.

Using implicit differentiation on (6) we obtain, after simplification,

$$(2aX + bY + d) + (bX + 2cY + e)Y' = 0.$$

If (x, y) is the closest point on the conic to the line $AX + BY + C - 0$, then we can replace Y' by $-A/B$ and assert (x, y) to satisfy the equation

$$2aX + bY + d + (bX + 2cY + e)(-A/B) = 0. \tag{10}$$

Provided one does not have

$$\left.\begin{array}{r} 2a - bA/B = 0 \\ b - 2cA/B = 0 \\ d - eA/B = 0 \end{array}\right\}, \tag{11}$$

equation (10) is the equation of a straight line and we can appeal to Theorem 3.2 to find (x, y) and whatever other point of intersection the line might have with the conic, drop the perpendicular from this (these) point(s) to the straight line and, after sorting things out[20], determine the distance from the line to the conic.

Should the conditions (11) hold, then

$$d = eA/B$$
$$b = 2cA/B$$
$$a = c(A/B)^2$$

and (6) successively becomes

$$c\frac{A^2}{B^2}X^2 + 2c\frac{A}{B}XY + cY^2 + e\frac{A}{B}X + eY + f = 0$$

$$c\left(\frac{A}{B}X + Y\right)^2 + e\left(Y + \frac{A}{B}X\right) + f = 0$$

$$c\left(Y + \frac{A}{B}X\right)^2 + e\left(Y + \frac{A}{B}X\right) + f = 0,$$

---

[19] In the case of an hyperbola, there can be two such points, one on either branch, should the line pass through the center.

[20] I.e. ruling out the intersecting point of maximum distance from the line in the case of an ellipse, or the merely nearest point on the more distant branch of an hyperbola.

and the quadratic formula yields

$$Y + \frac{A}{B}X = \frac{-e \pm \sqrt{e^2 - 4cf}}{2c}$$

and the conic is degenerate, which we have already ruled out.   □

As for the failure to be able to constructibly drop a normal from a constructible point to a constructibly determined conic section, i.e. to determine the distance from a point to the conic section, counterexamples abound. In the parabolic case, the failure in general is easily shown to be universal:

3.10 Theorem. *For any constructibly determined parabola there is a constructible point (α, β) such that the distance from the point to the parabola is not a constructible number, i.e. the normal from the point to the parabola cannot be constructed by means of ruler and compass.*

Proof. Via an affine transformation (i.e. one of the form, $X = mX_1 + nY_1 + p, Y = qX_1 + rY_1 + s$), every constructibly determined parabola can be put into the form $Y = aX^2$ for some constructible $a \neq 0$. The slope of the tangent line to the curve at a point (x, y) on the parabola is $y' = 2ax$, whence the slope of the normal at that point is $-1/(2ax)$, and, if this normal passes through the point (α, β), we have

$$-\frac{1}{2ax} = \frac{y - \beta}{x - \alpha},$$

whence (x, y) satisfies

$$-(X - \alpha) = 2aX(Y - \beta),$$

i.e.

$$-X + \alpha = 2aXY - 2a\beta X$$
$$= 2aX(aX^2) - 2a\beta X,$$

i.e.

$$2a^2 X^3 + (1 - 2a\beta)X - \alpha = 0. \tag{12}$$

If we now choose a to be an arbitrary nonzero constructible number, $\beta = 1/(2a)$, and $\alpha = 4a^2$, the equation becomes

$$2a^2 X^3 - 4a^2 = 0,$$

i.e. $X^3 - 2 = 0$, which we know has no constructible solution.

We are not quite done. We must verify that (α, β) does not lie on the curve. Should it happen to lie on the parabola, we would have $\beta = a\alpha^2$, i.e.

$$\frac{1}{2a} = a(4a^2)^2.$$

But then $2 = 4a^2(4a^2)^2 = (4a^2)^3$ and $4a^2$ is not constructible and thus neither is a, contrary to assumption.   □

The elliptic and hyperbolic cases are computationally more intense and we shall have to settle for a pair of examples. First, by way of preparation, assume $a, c \neq 0, a \neq c$ and consider the equation,

$$aX^2 + cY^2 = 1. \tag{13}$$

By implicit differentiation, at any point (x, y) on the curve, we have

$$2ax + 2cyy' = 0 \quad \text{whence} \quad y' = -\frac{ax}{cy}.$$

The slope of the normal at (x, y) is the negative reciprocal of y' and, should the normal pass through a point (α, β) not on the curve, we have

$$\frac{cy}{ax} = \frac{y - \beta}{x - \alpha},$$

so

$$\frac{y - \beta}{cy} = \frac{x - \alpha}{ax} = \text{ some constant } \lambda.$$

Note that, if $x, y, \alpha, \beta, a, c$ are constructible, then so is λ. Solving for x, y in terms of λ, we find

$$x = \frac{\alpha}{1 - a\lambda}, \qquad y = \frac{\beta}{1 - c\lambda}.$$

Because x, y satisfy (13) we have

$$a\left(\frac{\alpha}{1 - a\lambda}\right)^2 + c\left(\frac{\beta}{1 - c\lambda}\right)^2 = 1,$$

i.e.

$$a\alpha^2(1 - c\lambda)^2 + c\beta^2(1 - a\lambda)^2 = (1 - a\lambda)^2(1 - c\lambda)^2,$$

which is biquadratic in λ. Simplifying this, we have

$$a^2c^2\lambda^4 - 2ac(a + c)\lambda^3 + (a^2 + c^2 + 4ac - a\alpha^2c^2 - a^2\beta^2c)\lambda^2 \\ + 2(a\alpha^2c + a\beta^2c - a - c)\lambda + 1 - a\alpha^2 - c\beta^2 = 0. \tag{14}$$

3.11 Example. The distance from the constructible point $(1, 1)$ to the constructibly determined hyperbola $X^2 - Y^2 = 1$ is not constructible.

For this choice of a, c, α, β, (14) reads

$$\lambda^4 - 2\lambda^2 - 4\lambda + 1 = 0, \tag{15}$$

no solution λ of which is rational. The cubic resolvent of this polynomial is

$$X^3 - 4X^2 - 16 = 0 \tag{16}$$

which also has no rational roots.[21] Hence the cubic resolvent has no constructible roots and Theorem 2.14 tells us (15) doesn't either.

3.12 Example. The distance from the constructible point $(1,1)$ to the constructibly determined ellipse $X^2 + 2Y^2 = 1$ is not constructible.

For this choice of a, c, α, β, (14) reads

$$4\lambda^4 - 12\lambda^3 + 7\lambda^2 + 2\lambda - 2 = 0. \tag{17}$$

Dividing the coefficients by 4 and then making the substitution $\lambda = Y + 3/4$ results in

$$Y^4 - \frac{13}{8}Y^2 - \frac{1}{4}Y - \frac{23}{256} = 0,$$

i.e. $256Y^4 - 416Y^2 - 64Y - 23 = 0$. Setting $4Y = Z$, this becomes

$$Z^4 - 26Z^2 - 16Z - 23 = 0,$$

which has no rational solutions. The cubic resolvent of this is

$$X^3 - 52X^2 + 768X - 256, \tag{18}$$

The substitution $X = 4W$ results in

$$64W^3 - 832W^2 + 3072W - 256,$$

whence in

$$W^3 - 13W^2 + 48W - 4,$$

which is easily seen to have no rational roots. Hence (18) has no rational roots, whence Z, Y, and finally λ are not constructible.

4 Quintisection

After reading that the 60 degree angle cannot be trisected by ruler and compass alone, one might like to try one's hand at the 30 degree angle or the 120 degree angle. These cases can be given treatments similar to that of the 60 degree angle, and carrying out such a proof can help one to be sure one understands the original proof. However, these cases can also be reduced quickly to the 60 degree case. If, for example, we could trisect the 30 degree angle, we would have constructed a 10 degree angle, the doubling of which— easily

---

[21] This is verified, of course, by the old Gaussian observation that the rational roots of a polynomial with integral coefficients and leading coefficient 1 are divisors of the constant coefficient. Thus we need only test (15) for $\lambda = \pm 1$, and (16) for $\lambda = \pm 1, \pm 2, \pm 4, \pm 8, \pm 16$. I briefly discuss how to perform so many checks with a minimal amount of effort in the section on computation in the chapter on cubic equations.

accomplished by ruler and compass— would trisect the 60 degree angle. Hence it cannot be done. The 120 degree angle cannot be trisected by ruler and compass because its trisected angle of 40 degrees, when bisected— another easy ruler and compass construction—, would again trisect the 60 degree angle.

Recalling that Euclid gave a construction of the regular pentagon, the central angles of which are $360/5 = 72$ degrees, we can quickly characterise angles of an integral number of degrees that can be constructed by ruler and compass, whence we can characterise those such angles which can be trisected.

4.1 Theorem. *For positive integers n, the angle of n degrees can be constructed by ruler and compass iff 3 divides n.*

Proof. The 72 degree angle can be constructed. From a 45 and a 30 degree angle we obtain a 75 degree angle. Subtracting the 72 degree angle from it produces a 3 degree angle. If 3 divides n, say $n = 3k$, we can add k copies of a 3 degree angle to produce an n degree angle.

If n is not divisible by 3, it leaves a remainder of 1 or 2 after dividing by 3, whence the constructibility of an n degree angle would yield, in the face of what has just been proven, the constructibility of either a 1 or a 2 degree angle. However, through the use of multiple copies of such an angle, its constructibility would yield the constructibility of the forbidden 20 degree angle. Hence, if n is not divisible by 3 the angle of n degrees is not constructible by ruler and compass.   □

4.2 Corollary. *For positive integers n, the angle of n degrees is constructible and can be trisected iff 9 divides n.*

Because bisection can be accomplished by ruler and compass, the proof of the Theorem can be generalised to yield.

4.3 Theorem. *For positive integers n, k, the angle of $n/2^k$ degrees can be constructed by ruler and compass iff 3 divides n.*

The constructibility of the 72 degree angle can be established and the pentagon constructed in a number of different ways. Algebraically, we can follow the approach used to establish non-trisectability to "quintisect" the 360 degree angle. One starts with the trigonometric identity

$$\cos 5\theta = 16 \cos^5 \theta - 20 \cos^3 \theta + 5 \cos \theta,$$

which can again be established by repeated application of the Addition Formulæ for sines and cosines. Letting $a = \cos 5\theta$, this means considering the equation

$$16X^5 - 20X^3 + 5X - a = 0, \qquad -1 \le a \le 1.$$

Multiplying by 2,

$$32X^5 - 40X^3 + 10X - 2a = 0, \tag{19}$$

and substituting $Y = 2X$ yields the simpler looking equation:

$$Y^5 - 5Y^3 + 5Y - 2a = 0, \qquad -1 \le a \le 1. \tag{20}$$

4.4 Construction. To construct the 72 degree angle, we set $5\theta = 360$, whence $a = \cos 360 = 1$. Note that $5\theta = 0, 720, 1080$, and 1440 also have 1 as their cosine, whence the doubled cosines of their fifths,

$$2\cos 0 = 2, \quad 2\cos(360/5) = 2\cos(1440/5), \quad 2\cos(720/5) = 2\cos(1080/5), \tag{21}$$

i.e.

$$2\cos 0 = 2, \quad 2\cos 72 = 2\cos 288, \quad 2\cos 144 = 2\cos 216,$$

all satisfy the equation

$$Y^5 - 5Y^3 + 5Y - 2 = 0.$$

Because $Y = 2$ is a root, we can factor out $Y - 2$:

$$(Y - 2)(Y^4 + 2Y^3 - Y^2 - 2Y + 1) = 0.$$

Thus, we consider the equation

$$Y^4 + 2Y^3 - Y^2 - 2Y + 1 = 0. \tag{22}$$

There are two ways we can go about solving this. We can note the symmetry of the coefficients about the central term and mimic the nontrisectability proof by dividing by Y^2 and rearranging the terms:

$$Y^2 + \frac{1}{Y^2} + 2Y - \frac{2}{Y} - 1 = 0.$$

Substituting $Z = Y - \frac{1}{Y}, Z^2 = Y^2 + \frac{1}{Y^2} - 2$, yields the quadratic equation

$$Z^2 + 2 + 2Z - 1 = 0,$$

i.e.

$$Z^2 + 2Z + 1 = 0,$$

which is a perfect square $(Z + 1)^2 = 0$. Hence $Z = -1$.

Recalling $Z = Y - \frac{1}{Y}$, this yields successively

$$Y - \frac{1}{Y} = -1$$
$$Y^2 - 1 = -Y$$
$$Y^2 + Y - 1 = 0,$$

and the quadratic formula yields

$$Y = \frac{-1 \pm \sqrt{1 + 4}}{2} = \frac{-1 \pm \sqrt{5}}{2}$$

and since $Y = 2\cos\theta$, this means

$$\cos\theta = \frac{-1 \pm \sqrt{5}}{4},$$

both numbers being constructible. Observing which angles lie in which quadrants sorts things out:

$$\cos 72 = \cos 288 = \frac{-1 + \sqrt{5}}{4}$$

$$\cos 144 = \cos 216 = \frac{-1 - \sqrt{5}}{4}.$$

A second approach to solving equation (22) algebraically is to note that its roots (21) are repeated roots. (The angles do not repeat, but their cosines do.) This suggests the polynomial (22) is a perfect square and should be of the form $(Y^2 + bY \pm 1)^2$. Assume this to be the case and solve for b:

$$Y^4 + 2Y^3 - Y^2 - 2Y + 1 = (Y^2 + bY \pm 1)^2$$
$$= Y^4 + 2bY^3 + (\pm 2 + b^2)Y^2 \pm 2bY + 1.$$

From the coefficients of the cubic terms, we see $2b = 2$, i.e. $b = 1$. Each of the quadratic and linear terms tells us that the sign of 1 is negative. Thus (22) becomes

$$(Y^2 + Y - 1)^2 = 0,$$

i.e. $Y^2 + Y - 1 = 0$ and again

$$Y = \frac{-1 \pm \sqrt{5}}{2}.$$

4.5 Construction. Quintisecting any angle along these lines will construct a 72 degree angle. For, if we can quintisect 5θ by these algebraic means, we can do the same to $5\theta + 360$ to obtain $\theta + 72$. Subtracting the smaller from the larger gives a 72 degree angle. That said, it must be admitted that not all constructions are created equal. To demonstrate this, we consider $\cos 5\theta = -1/2$, i.e. $5\theta = 120$ or 240, or $120 + 360k$ or $240 + 360k$ for $k = 1, 2, 3, 4$. The angles θ are thus $24 + 72k$ and $48 + 72k$ for $k = 0, \ldots, 4$ and, being divisible by 3, are all constructible by ruler and compass. Let us pretend we don't know this and approach the problem algebraically.

Equation (20) becomes

$$Y^5 - 5Y^3 + 5Y + 1 = 0. \tag{23}$$

This has $-1 \left(= 2\cos\left(\frac{120 + 3 \cdot 360}{5}\right) = 2\cos 240\right)$ as a root, whence the factorisation,

$$(Y + 1)(Y^4 - Y^3 - 4Y^2 + 4Y + 1),$$

and we must consider the equation

$$Y^4 - Y^3 - 4Y^2 + 4Y + 1 = 0. \tag{24}$$

The substitution $Z = Y - 1$, i.e. $Y = Z + 1$ (trial and error) will make this more manageable:

$$
\begin{array}{rl}
(Z+1)^4 &= Z^4 + 4Z^3 + 6X^2 + 4Z + 1 \\
-(Z+1)^3 &= \quad\quad -Z^3 - 3Z^2 - 3Z - 1 \\
-4(Z+1)^2 &= \quad\quad\quad\quad - 4Z^2 - 8Z - 4 \\
+4(Z+1) &= \quad\quad\quad\quad\quad\quad 4Z + 4 \\
+1 &= \quad\quad\quad\quad\quad\quad\quad\quad + 1 \\
\hline
& Z^4 + 3Z^3 - \quad Z^2 - 3Z + 1 = 0.
\end{array}
$$

The new coefficients are more symmetric, whence we once again divide by Z^2 and rearrange terms:

$$Z^2 + \frac{1}{Z^2} + 3\left(Z - \frac{1}{Z}\right) - 1 = 0.$$

Setting $W = Z - \frac{1}{Z}, W^2 = Z^2 + \frac{1}{Z^2} - 2$, this becomes

$$W^2 + 2 + 3W - 1 = 0,$$

i.e. $W^2 + 3W + 1 = 0$. The quadratic formula yields

$$W = \frac{-3 \pm \sqrt{9-4}}{2} = \frac{-3 \pm \sqrt{5}}{2}.$$

Again,

$$Z - \frac{1}{Z} = \frac{-3 \pm \sqrt{5}}{2}$$

$$Z^2 - 1 = \frac{-3 \pm \sqrt{5}}{2} Z$$

$$Z^2 + \frac{3 \mp \sqrt{5}}{2} Z - 1 = 0.$$

Using the two values of the signs and applying the quadratic formula gives 4 values of Z:

$$
Z = \begin{cases}
\dfrac{-\frac{3+\sqrt{5}}{2} \pm \sqrt{\frac{9+5+6\sqrt{5}}{4} + 4}}{2} = \dfrac{-3 - \sqrt{5} \pm \sqrt{30 + 6\sqrt{5}}}{4} \\[3ex]
\dfrac{-\frac{3-\sqrt{5}}{2} \pm \sqrt{\frac{9+5-6\sqrt{5}}{4} + 4}}{2} = \dfrac{-3 + \sqrt{5} \pm \sqrt{30 - 6\sqrt{5}}}{4}.
\end{cases}
$$

Recalling that $\cos\theta = Y/2 = (Z+1)/2$, we have

$$\cos 24 = \frac{1 + \sqrt{5} + \sqrt{30 - 6\sqrt{5}}}{8} = \cos 336$$

$$\cos 48 = \frac{1 - \sqrt{5} + \sqrt{30 + 6\sqrt{5}}}{8} = \cos 312$$

$$\cos 96 = \frac{1 + \sqrt{5} - \sqrt{30 - 6\sqrt{5}}}{8} = \cos 264$$

$$\cos 168 = \frac{1 - \sqrt{5} - \sqrt{30 + 6\sqrt{5}}}{8} = \cos 192,$$

and, of course, the "trivial" solution to (23):

$$\cos 240 = -\frac{1}{2} = \cos 120.$$

It is not hard to see that none of $\cos 24, \cos 48, \cos 96$, and $\cos 148$ is rational. (Indeed, Euclid devoted much of Book X of the *Elements* to proving the irrationality of numbers of this form.) That none of them are written in the form $p + q\sqrt{n}$ for p, q, n rational suggests that the polynomial in (24) cannot be factored over the rationals. Indeed it cannot. As none of the roots is rational, any factorisation over the rationals must be into quadratic factors of the form

$$(X - \alpha)(X - \beta) = X^2 - (\alpha + \beta)X + \alpha\beta,$$

for two of the roots α, β given above. For such a pairing, $\alpha + \beta$ must be rational.

Without loss of generality, we may assume $\alpha = \cos 24$ and consider its pairings with the three remaining values of β:

$$\cos 24 + \cos 96 = \frac{1 + \sqrt{5}}{4}$$

$$\cos 24 + \cos 48 = \frac{2 + \sqrt{30 - 6\sqrt{5}} + \sqrt{30 + 6\sqrt{5}}}{8}$$

$$\cos 24 + \cos 168 = \frac{2 + \sqrt{30 - 6\sqrt{5}} - \sqrt{30 + 6\sqrt{5}}}{8}.$$

The first of these is obviously irrational and the latter two are rational or irrational as is $\sqrt{30 - 6\sqrt{5}} \pm \sqrt{30 + 6\sqrt{5}}$. Thus, let

$$q = \sqrt{30 - 6\sqrt{5}} + \sqrt{30 + 6\sqrt{5}}$$

be assumed rational. Then

$$q^2 = 30 - 6\sqrt{5} + 2\sqrt{30 - 6\sqrt{5}}\sqrt{30 + 6\sqrt{5}} + 30 + 6\sqrt{5}$$

$$= 60 + 2\sqrt{30^2 - 6^2 \cdot 5} = 60 + 2 \cdot 6\sqrt{25 - 5}$$

$$= 60 + 12\sqrt{20} = 60 + 24\sqrt{5}$$

is rational, whence so too is $\sqrt{5} = (q^2 - 60)/24$. But this contradicts the irrationality of $\sqrt{5}$ and we conclude $\cos 24 + \cos 48$ is not rational. Similarly, one can show that $\cos 24 + \cos 168$ is irrational and conclude that the polynomial in (24) cannot be factored over $\mathbb{Q}$.

4.6 Construction. Algebraically, the easiest quintisection may be had by choosing $\cos \theta = 0$, i.e. $5\theta = 90$ or 270 (both $+ 360k$ for $k = 1, 2, 3, 4$). In this case (20) becomes

$$Y^5 - 5Y^3 + 5Y = 0.$$

Factoring out Y $(0 = \cos\left(\frac{90 + 360}{5}\right))$ yields

$$Y^4 - 5Y^2 + 5 = 0, \tag{25}$$

which is quadratic in Y^2:

$$Y^2 = \frac{5 \pm \sqrt{25 - 4 \cdot 5}}{2} = \frac{5 \pm \sqrt{5}}{2}$$

$$Y = \begin{cases} \pm\sqrt{\dfrac{5 + \sqrt{5}}{2}} \\ \pm\sqrt{\dfrac{5 - \sqrt{5}}{2}}. \end{cases}$$

Again, $\cos \theta = Y/2$ and, sorting out the angles, we have

$$\cos 18 = \frac{1}{2}\sqrt{\frac{5 + \sqrt{5}}{2}} = \cos 342$$

$$\cos 162 = -\frac{1}{2}\sqrt{\frac{5 + \sqrt{5}}{2}} = \cos 198$$

$$\cos 234 = -\frac{1}{2}\sqrt{\frac{5 - \sqrt{5}}{2}} = \cos 120$$

$$\cos 306 = \frac{1}{2}\sqrt{\frac{5 - \sqrt{5}}{2}} = \cos 54,$$

and, of course, $\cos 90 = 0 = \cos 270$.

Polynomial (25) does not factor, as we can probably prove by the method used in Construction 4.5. However, this is a prime candidate for an application of a result of abstract algebra:

4.7 Theorem (Eisenstein's Criterion). *Let $P(X) = a_n X^n + \ldots + a_1 X + a_0$ be a polynomial with integral coefficients, with $a_n \neq 0$. If p is a prime number satisfying either of the two sets of conditions,*
(c_1) *p divides $a_1, \ldots, a_n$, p does not divide a_0, p^2 does not divide a_n,*
(c_2) *p divides $a_0, \ldots, a_{n-1}$, p does not divide a_n, p^2 does not divide a_0,*
then P does not factor over $\mathbb{Q}$.

Eisenstein's Criterion applies immediately to (25). It also applies to (24) if we first make the substitution $Y = Z - 1$ to obtain $Z^4 - 5Z^3 + 5Z^2 + 5Z - 5 = 0$.

The failure of the polynomials of equations (24) and (25) of Constructions 4.5 and 4.6 to factor can be combined with a little field theory to show that the complicated expressions for the roots are necessarily more complicated than those of Construction 4.4. Construction 4.4 puts $\cos 72$ into $\mathbb{Q}[\sqrt{5}]$; Construction 4.5 puts $\cos 24$ into $\mathbb{Q}[\sqrt{5}][\sqrt{30 - 6\sqrt{5}}]$; and Construction 4.6 puts $\cos 18$ into $\mathbb{Q}[\sqrt{5}][\sqrt{(5 + \sqrt{5})/2}]$. These latter two chains cannot be shortened. And one can show $\cos 9 (= \cos(45/5))$ requires a chain of three extensions of $\mathbb{Q}$. Attempting to solve the corresponding polynomial in this case is a bit hairy, and I recommend a different approach. One possibility is to use some trigonometric identities. For example, $36 = \frac{1}{2} \times 72$, whence

$$\cos 36 = \sqrt{\frac{1 + \cos 72}{2}} = \sqrt{\frac{4 - 1 + \sqrt{5}}{4 \cdot 2}}$$

$$= \sqrt{\frac{3 + \sqrt{5}}{8}} = \frac{1}{2}\sqrt{\frac{3 + \sqrt{5}}{2}}$$

$$\sin 36 = \sqrt{1 - \cos^2 36} = \frac{1}{2}\sqrt{\frac{5 - \sqrt{5}}{2}}.$$

But,

$$\cos 9 = \cos(45 - 36) = \cos 45 \cos 36 + \sin 45 \sin 36$$

$$= \frac{\sqrt{2}}{2}\left(\frac{1}{2}\sqrt{\frac{3 + \sqrt{5}}{2}} + \frac{1}{2}\sqrt{\frac{5 - \sqrt{5}}{2}}\right)$$

$$= \frac{\sqrt{3 + \sqrt{5}} + \sqrt{5 - \sqrt{5}}}{4},$$

or, more simply,

$$\cos 9 = \sqrt{\frac{1 + \cos 18}{2}} = \cdots = \frac{1}{2}\sqrt{\frac{4 + \sqrt{10 + 2\sqrt{5}}}{2}}.$$

Perhaps more interesting is the application of Eisenstein's Criterion and some field theory to impossibility proofs:

4.8 Construction. Let $\cos 5\theta = 1/4$. Whatever 5θ may be, it is clear that the angle is constructible by ruler and compass. But can θ be so constructed? The equation in this case is

$$Y^5 - 5Y^3 + 5Y - \frac{1}{2} = 0,$$

or

$$2Y^5 - 10Y^3 + 10Y - 1 = 0.$$

Using 2 as our prime, Eisenstein's Criterion tells us this cannot be factored. Hence any number field extending $\mathbb{Q}$ containing a root of this equation will have a degree divisible by 5 and will not be a power of 2 as is the case with a field constructible by a finite chain of quadratic extensions of number fields.[22] Hence, $\cos\theta$ is not a constructible number and the angle θ cannot be constructed by ruler and compass.

We can generalise this:

4.9 Theorem. *A constructible angle with rational cosine $\frac{m}{n}$, with m, n relatively prime cannot be quintisected by means of ruler and compass in the following cases:*
a. $n = 4k$, where k is odd
b. $n = pk$, where p is an odd prime and p does not divide k
c. $m = 5k$, where 5 does not divide k.

For, in each of these cases, the equation,

$$Y^5 - 5Y^3 + 5Y - 2\frac{m}{n} = 0$$

transforms into

$$nY^5 - 5nY^3 + 5nY - 2m = 0$$

and Eisenstein's Criterion applies.

5 Algebraic Numbers

The crucial steps involved in the nonconstructibility proofs of the previous section were Eisenstein's Criterion and some unproven assertions about the degrees of polynomials with rational coefficients that constructible numbers can be zeroes of. Neither of these results is all that difficult. To apply them, Eisenstein's Criterion needs no explanation, but the latter does.

5.1 Definition. *Let F be a number field. A number $\alpha \in \mathbb{R}$ is algebraic over F if there is a polynomial*

$$P(X) = a_n X^n + \ldots + a_1 X + a_0$$

with coefficients $a_i \in F, a_n \neq 0, n > 0$ such that $P(\alpha) = 0$. α is algebraic if it is algebraic over $\mathbb{Q}$.

---

[22] A detailed proof of this assertion is the subject of the next section.

5.2 Definition. *Let α be algebraic over F. A minimal polynomial for α is a non-constant polynomial P with coefficients in F such that $P(\alpha) = 0$ is of minimal degree, i.e. if $P(\alpha) = 0$ and $Q(\alpha) = 0$, then the degree of Q, $\deg(Q)$ is at least that of P. The degree of such P is called the* degree of α.

5.3 Lemma. *Let α be algebraic over F with minimal polynomial $P(X)$ and suppose $Q(X)$ is another polynomial over F for which $Q(\alpha) = 0$. Then: P divides Q.*

Proof. Apply long division to obtain

$$Q(X) = P(X)Q_1(X) + R(X),$$

where Q_1, R are polynomials over F with the degree of R being strictly less than that of P. This can be done over F because the division procedure for polynomials requires the arithmetic operations be applied to the coefficients of the polynomials and F is closed under these, i.e. the polynomials resulting from such a division when performed over the reals have their coefficients in F. But

$$0 = Q(\alpha) = P(\alpha)Q_1(\alpha) + R(\alpha) = 0 + R(\alpha) = R(\alpha),$$

and by the minimality of P, R must be the constant 0, i.e. $Q(X) = P(X)Q_1(X)$. $\qquad\square$

5.4 Corollary. *Let α be algebraic over a number field F with minimal polynomial P. Then P is* prime: *if $Q(X), R(X)$ are polynomials over F for which $P(X)$ divides $Q(X)R(X)$, then P divides one of Q and R.*

Proof. Let $Q(X)R(X) = P(X)D(X)$. Then

$$Q(\alpha)R(\alpha) = P(\alpha)D(\alpha) = 0$$

and one of $Q(\alpha)$ and $R(\alpha)$ is 0. $\qquad\square$

Along the lines of the Lemma and its Corollary is another result that should be mentioned.

5.5 Lemma. *Let α be algebraic over a number field F with minimal polynomial P. Then P does not factor over F into two polynomials of lower degree.*

For, if P factored, one of the factors would map α to 0 contradicting the minimality of P.

5.6 Theorem. *Let F be a number field, $\alpha \in F$, $\sqrt{\alpha} \notin F$. If β is algebraic over $F[\sqrt{\alpha}]$, then β is algebraic over F. Moreover, if the degree of β over $F[\sqrt{\alpha}]$ is n, its degree over F is either n or $2n$.*

Proof. Let $P(X) = a_n X^n + \ldots + a_1 X + a_0$ with $a_n, \ldots, a_0 \in F[\sqrt{\alpha}]$ be such that $P(\beta) = 0$. We write each a_i in the form $b_i + c_i\sqrt{\alpha}$, with $b_i, c_i \in F$. Thus,

$$P(X) = (b_n + c_n\sqrt{\alpha})X^n + \ldots + (b_1 + c_1\sqrt{\alpha})X + (b_0 + c_0\sqrt{\alpha})$$
$$= b_nX^n + \ldots + b_1X + b_0 + \sqrt{\alpha}(c_nX^n + \ldots + c_1X + c_0)$$
$$= Q(X) + \sqrt{\alpha}\,R(X),$$

where Q, R have coefficients in F.

From $P(\beta) = 0$ follows

$$Q(\beta) + \sqrt{\alpha}\,R(\beta) = 0$$
$$Q(\beta) = -\sqrt{\alpha}\,R(\beta)$$
$$Q(\beta)^2 = \alpha R(\beta)^2$$
$$Q(\beta)^2 - \alpha R(\beta)^2 = 0,$$

i.e. β is a root of the polynomial $Q(X)^2 - \alpha R(X)^2$ with coefficients in F. Thus β is algebraic over F.

Determining the degree of β over F from that over $F[\sqrt{\alpha}]$ is where the real work lies and was the reason for proving the lemmas.

Let P, Q, R be as above, only now assume P is the minimal polynomial for β over $F[\sqrt{\alpha}]$. Let $\overline{P}(X) = Q(X) - \sqrt{\alpha}\,R(X)$ and consider the factorisation:

$$Q(X)^2 - \alpha R(X)^2 = (Q(X) + \sqrt{\alpha}\,R(X))(Q(X) - \sqrt{\alpha}\,R(X))$$
$$= P(X) \cdot \overline{P}(X).$$

If T is the minimal polynomial for β over F, then T divides $Q(X)^2 - \alpha R(X)^2$ over F, whence also over $F[\sqrt{\alpha}]$,

$$Q(X)^2 - \alpha R(X)^2 = T(X) \cdot D(X),$$

for some D with coefficients in F. Thus

$$P(X) \cdot \overline{P}(X) = T(X) \cdot D(X).$$

Now P divides T by Lemma 5.3, whence

$$\deg(T) \geq \deg(P) = n. \tag{26}$$

The question is: what does $\overline{P}$ do?

$P(X)$ is prime by Corollary 5.4 because it is the minimal polynomial of β over $F[\sqrt{\alpha}]$. $\overline{P}(X)$ is also prime— because $P(X)$ is. To see this, define more generally the conjugate $\overline{S}(X)$ of any polynomial $S(X)$ with coefficients in $F[\sqrt{\alpha}]$ by replacing every occurrence of $\sqrt{\alpha}$ in a coefficient of S by $-\sqrt{\alpha}$. It is easy to verify that

$$\overline{S_1(X) \cdot S_2(X)} = \overline{S_1}(X) \cdot \overline{S_2}(X)$$
$$\overline{\overline{S(X)}} = S(X).$$

Thus, if $\overline{P}$ divides a product $S_1 \cdot S_2$, we have

$$S_1(X) \cdot S_2(X) = \overline{P(X)} \cdot U(X)$$

$$\overline{S_1(X) \cdot S_2(X)} = \overline{\overline{P(X)} \cdot U(X)}$$

$$\overline{S_1(X)} \cdot \overline{S_2(X)} = \overline{\overline{P(X)}} \cdot \overline{U(X)}$$

$$\overline{S_1(X)} \cdot \overline{S_2(X)} = P(X) \cdot \overline{U(X)}$$

and $P(X)$ divides one of $\overline{S_1(X)}, \overline{S_2(X)}$. But then $\overline{P}$ divides one of $\overline{\overline{S_1}}, \overline{\overline{S_2}}$, i.e. one of S_1, S_2.

So $\overline{P}$ is prime and must divide T or D.

If $\overline{P}$ divides T, then $P \cdot \overline{P}$ divides T and

$$2n = \deg(P) + \deg(\overline{P}) \leq \deg(T) \leq \deg(TD) = \deg(P\overline{P}) = 2n,$$

i.e. $\deg(T) = 2n$.

If $\overline{P}$ divides D,

$$\deg(T) = \deg(TD) - \deg(D)$$
$$\leq 2n - n, \text{ because } \deg(D) \geq n$$
$$\leq n.$$

But $\deg(P) \geq n$, by (26), whence $\deg(T) = n$.   □

5.7 Corollary. *Every constructible real number is algebraic of degree 2^n for some $n \geq 0$.*

Proof. If β is constructible, there is a chain

$$\mathbb{Q} = F_0 \subseteq F_1 \subseteq \ldots \subseteq F_m, \ \beta \in F_m$$

of number fields, each F_{i+1} extending F_i by the adjunction of a square root. Since $\beta \in F_m$, it satisfies the equation $X - \beta = 0$ with coefficients in F_m, i.e. β is algebraic of degree $1 = 2^0$ over F_m. By the Theorem, β is algebraic over F_{m-1}, whence over $F_{m-2}, \ldots$, whence over $F_0 = \mathbb{Q}$. Moreverover, the degree of β over each F_{m-i-1} is either the same as that over F_{m-i} or double it.   □

This Corollary, particularly the fact that the degree of a constructible real number is a power of 2, justifies our nonconstructibility assertions made at the end of the preceding section. In particular, for $\beta = \cos\theta$ for some angle θ for which $\cos 5\theta = \frac{1}{4}$, we know that β is the root of an irreducible polynomial of degree 5. If Q is the minimal polynomial of β, then Q divides P, whence the degree of Q is that of P (i.e. P is a constant multiple of Q). Thus β has degree 5 and is not constructible.

6 Petersen Revisited

The elementary impossibility proofs given in sections 1 - 3 were later simpli-
fications. The original proofs by Wantzel, Petersen, Klein, and Pierpont all
proceeded along the lines of section 5: the irreducible equations satisfied by
constructible reals all have degrees that are powers of 2; the solutions to the
problems considered satisfy irreducible polynomials of degrees that are not
powers of 2; therefore the solutions are not constructible. Examples 3.11 and
3.12 are exceptions. They resulted in equations of degree 4 that had no constructible solutions, which fact could not be proven by appeal to the degree
alone. For these we had to appeal to a special reduction to the cubic case.

    Petersen proves his general results, given in opening statements 1 - 4 of
section 3 by arguing along similar lines. First he proves statement 3 by assuming the curve $P(X, Y) = 0$ to intersect all constructible lines of a given pencil
of lines only in constructible points. He then argues as in section 5 that the
degree of P is a power of 2. The rest of his proof remains a bit of a mystery to
me.[23] At the cost of relying on a deeper, but moderately well-known, result,
I can present a somewhat less intricate proof than Petersen did. This deeper
result is known as the Hilbert Irreducibility Theorem and concerns irreducible
polynomials in two variables.

    The study of polynomials in two variables is more complicated than that
of polynomials in a single variable. For example, the division algorithm fails:
the total degree of $X^2Y + 1$ is 3, less than the degree 4 of $XY^3 + 1$, yet one
cannot write

$$XY^3 + 1 = (X^2Y + 1) \cdot Q(X, Y) + R(X, Y)$$

with R of degree less than 3. Some familiar properties, however, do carry over.

6.1 Lemma. *For any polynomials $P(X, Y)$ and $Q(X, Y)$,*

$$\deg(P \cdot Q) = \deg(P) + \deg(Q).$$

6.2 Lemma. *Let F be a number field, $P(X, Y)$ a polynomial with coefficients
in F, and suppose P is irreducible over F, i.e. P does not factor over F into
polynomials of lower degree. Then: P is prime in the ring of polynomials over
F in two variables.*

    The first of these lemmas is not entirely trivial[24], but it is not difficult
either. The second is a bit deeper. First one shows that the collection $F[X]$
of polynomials in one variable over F has unique factorisation, and then that
this entails unique factorisation in the collection $F[X, Y] = (F[X])[Y]$ of

---

[23] In defense of my mathematical skills, I note that the adjective used in the *Dictionary of Scientific Biography* to describe his textbooks is "terse".

[24] Consider $(X + Y)(X - Y) = X^2 + XY - XY - Y^2$ in which the product of terms
of highest degrees in the factors can cancel out in the full product.

polynomials in two variables over F. The prime factorisation of an irreducible polynomial then tells us the irreducible polynomial is a prime.

I shall not prove these lemmas here, but shall take them as given. They occur in some undergraduate algebra textbooks and are not to be found in others. Thus we have already gone beyond what we can reasonably expect the student in one's history class to know, but I think not beyond what such a student could understand, find plausible, and be willing to accept on authority.

When it comes to irreducibility, the theory of polynomials in two variables is quite different from the theory in one variable, but there are easy results:

6.3 Example. Let $P(X, Y) = Y - P_1(X)$ where P_1 is a polynomial in X with real coefficients. P is irreducible over $\mathbb{R}$.

Proof. Suppose P factors into $Q(X, Y) \cdot R(X, Y)$, with Q, R having degrees less than the degree of P. Write

$$Q(X, Y) = \sum_{i=0}^{m} Q_i(X)Y^i, \qquad R(X, Y) = \sum_{j=0}^{n} R_j(X)Y^j,$$

with Q_m, R_n not identically 0. The term of highest degree in $P = Q \cdot R$, viewed as a polynomial in Y over $\mathbb{R}[X]$, is $Q_m(X)R_n(X)Y^{m+n}$ and $Q_m(X)R_n(X)$ is not identically 0. Comparing coefficients with those of P, we conclude $m + n = 1$, i.e. one of Q, R has Y occurring to at most degree 1 and the other has no Y at all. Without loss of generality, we may assume Y occurs in Q:

$$Q(X, Y) = Q_0(X) + cY, c \text{ a constant}$$
$$R(X, Y) = R_0(X).$$

Then
$$P(X, Y) = Y - P_1(X) = Q_0(X)R_0(X) + cR_0(X)Y,$$

and, comparing coefficients again, we see $R_0(X)$ is constant, i.e. $\deg(R) = 0$. Thus
$$\deg(P) = \deg(Q) + \deg(R) = \deg(Q),$$

contrary to assumption. Thus P does not factor over $\mathbb{R}$.   □

We can also demonstrate irreducibility through the observation that the graph of an equation $P(X, Y) = 0$ is the union of the graphs of the equations $Q(X, Y) = 0$ and $R(X, Y) = 0$ when $P(X, Y) = Q(X, Y) \cdot R(X, Y)$.

6.4 Example. Let $P(X, Y) = X^4 + Y^4 - 1$. P is irreducible over $\mathbb{R}$.

Proof. Suppose to the contrary that $P(X, Y) = Q(X, Y) \cdot R(X, Y)$, with Q, R each of degree less than 4. Now the graph of $P(X, Y) = 0$ is a bounded subset of the plane. Hence neither Q nor R can be linear, as the graph of a straight line is unbounded. This means both Q and R are quadratic, whence their graphs are conic sections. Again, the graphs being bounded in the plane, they are neither parabolic nor hyperbolic and they must in fact be ellipses. As the graph $P(X,Y) = 0$ is a simple closed curve, it thus must be an ellipse.

Now the curve $P(X,Y) = 0$ exhibits too much symmetry[25] to be an ordinary ellipse. If it is one, it must be a circle centered at the origin. But the points $(1,0)$ and $(1/\sqrt[4]{2}, 1/\sqrt[4]{2})$ lie on the curve at distances $1, \sqrt[4]{2}$, respectively, from the origin and we have a contradiction.   □

The observation that, from

$$P(X,Y) = Q(X,Y) \cdot R(X,Y)$$

follow

$$P(a,Y) = Q(a,Y) \cdot R(a,Y)$$

$$P(X,b) = Q(X,b) \cdot R(X,b)$$

for any $a, b \in F$, suggests a possible link between reducibility/irreducibility in $F[X,Y]$ and reducibility/irreducibility in $F[X]$ or $F[Y]$. There is indeed such a link, but it is not as strong as one would like.

6.5 Example. Let $P(X,Y) = (X^2 + 1)(Y^2 + 1)$. Then P is reducible over $\mathbb{Q}$, while, for every $a, b \in \mathbb{R}, P(a,Y)$ and $P(X,b)$ are irreducible over $\mathbb{R}$.

6.6 Example. Let $P(X,Y) = Y - X^3$. Then P is irreducible over $\mathbb{R}$ and

$$P(X,b) = b - X^3 = (\sqrt[3]{b} - X)(\sqrt[3]{b^2} + \sqrt[3]{b}X + X^2)$$

is reducible over $\mathbb{R}$ for all $b \in \mathbb{R}$.

The first of these examples is utterly devastating, the second less so. For, in the second, note that the factorisation only works when the cube root of b is available. If one looks at the situation over $\mathbb{Q}, P(X,Y)$ is irreducible and so is $P(X,b)$ for any $b \in \mathbb{Q}$ which is not a perfect cube.

6.7 Definition. *A field F of constructible real numbers is finitely generated if there is a chain $\mathbb{Q} = F_0 \subseteq F_1 \ldots \subseteq F_n = F$ of fields of the form $F_{i+1} = F_i[\sqrt{\alpha_i}]$, with $\alpha_i \in F_i$, i.e. if F can be generated over $\mathbb{Q}$ by the successive adjunction of finitely many square roots.*

6.8 Theorem (Hilbert Irreducibility Theorem). *Let F be a finitely generated field of constructible real numbers, let $P(X,Y)$ be a polynomial with coefficients in F, and suppose P is irreducible over F. Then: for any real numbers $\alpha < \beta$, there is a number $b \in F$ with $\alpha < b < \beta$ such that $P(X,b)$ is irreducible over F.*

---

[25] $P(-X,Y) = P(X,Y) = P(X,-Y)$, whence the X- and Y-axes are axes of symmetry. Moreover, $P(X,Y) = P(Y,X) = P(-Y,X)$, whence one also has symmetry with respect to the lines $Y = \pm X$. The centre of all this symmetry is the origin.

This is a strong version of the existence result: not only does there exist b for which $P(X, b)$ is irreducible, but the set of such is dense in F under the ordering inherited from $\mathbb{R}$. The proof, even without this extra information, is a bit deep.[26] Traditional proofs rely on techniques of the Calculus that lie just a little beyond what is covered in the standard American courses on the subject. Since the 1960s, the use of Calculus can be bypassed using tools of mathematical logic. Either way, the proof lies beyond the scope of this book, but the result ought not to be implausible to students in the history class— especially if one emphasises the nature of counterexamples over $\mathbb{R}$ and how they depend on $\mathbb{R}$ being almost algebraically closed.

We are now about ready to prove Petersen's third claim cited at the beginning of section 3. Before we do so, however, we must accurately state the result. I am inclined to think the word "coincide" in the statement of the result to be a bit too strong and one should read "occur" instead:

3' *If we are able by straight edge and compasses to determine the points, in which any line of a pencil of lines intersects a curve, (which does not pass through the vertex of the pencil), the order of this curve must be a power of 2 and there will be at least two lines in the pencil, of which the points of intersection with the curve occur in pairs.*

Formally this reads:

6.9 Theorem. *Let $P(X, Y)$ have coefficients in a finitely generated constructible number field F, P irreducible over F. Suppose for some point (α, β) not on the curve $P(X, Y) = 0$, the point(s) of the intersection of every line passing through (α, β) with the curve is (are) constructible. Then:*
i. the degree of P is a power of 2;
ii. for infinitely many lines passing through (α, β) the points of intersection with the curve $P(X, Y) = 0$ occur in pairs.

Proof. Let $P(X, Y)$ be given with coefficients in a finitely generated constructible number field F, suppose P is irreducible over F, and suppose further that (α, β) is a point not on the curve of P and for which every line passing through (α, β) intersects the curve only in constructible points. Changing variables if necessary ($U = X - \alpha, V = Y - \beta$), we may assume $(\alpha, \beta) = (0, 0)$, i.e. $P(0, 0) \neq 0$.

Any line passing through $(0, 0)$ other than the vertical one has the equation $Y = mX$ for some real number m. The points of intersection of $P(X, Y) = 0$ and $Y = mX$ are determined by the solutions to $P(X, mX) = 0$. Thus we consider the auxiliary polynomial $P(X, MX)$.

Claim. $P(X, MX)$, viewed as a polynomial in the variables X, M, is irreducible over F.

---

[26] The most accessible proof, for $F = \mathbb{Q}$, can be found in Hadlock's book cited in footnote 10.

Proof. Suppose to the contrary that $P(X, MX)$ factors into $Q(M, X) \cdot R(M, X)$ over F. Now it could happen that Q or R has terms $aM^i X^j$ with $j < i$ and when we try to replace $M^i X^j$ by $Y^i X^{j-i}$ that $j - i$ will be negative. Making this exchange, i.e. writing $M = Y/X$ and finding the common denominators, we see that there are polynomials $\widetilde{Q}, \widetilde{R}$ with coefficients in F and non-negative integers m, n such that

$$Q(M, X) = \frac{\widetilde{Q}(X, Y)}{X^m}, \qquad R(M, X) = \frac{\widetilde{R}(X, Y)}{X^n}.$$

Thus,

$$X^{m+n} P(X, Y) = \widetilde{Q}(X, Y) \cdot \widetilde{R}(X, Y).$$

Now P is irreducible over F, whence P is prime and it must divide one of $\widetilde{Q}$ or $\widetilde{R}$. Without loss of generality, we can assume P divides $\widetilde{Q}$. Then, up to a constant multiple, $\widetilde{R}$ is a power of X.

If $\deg(\widetilde{R}) > n$, then R is a power of X and

$$P(0, 0) = Q(M, 0) \cdot R(M, 0) = Q(M, 0) \cdot 0 = 0,$$

contrary to assumption. If $\deg(\widetilde{R}) < n$, then R is not a polynomial. Thus we conclude that $\deg(\widetilde{R}) = n$ and $R(M, X)$ is a constant. Thus the factorisation of $P(X, MX)$ was not into two nonconstant polynomials, i.e. $P(X, MX)$ is irreducible over F.   □

Getting back to the proof of the Theorem, we now wish to appeal to Hilbert Irreducibility Theorem to conclude that the polynomial $P(X, mX)$ is irreducible for some m. So long as the graph of $P(X, Y) = 0$ is not a straight line passing through $(0, 0)$, there will be some non-empty interval (a, b) such that every line $Y = mX$ for $m \in (a, b)$ has an intersection with the graph of $P(X, Y) = 0$. For infinitely many of these values of m, the polynomial $P(X, mX)$ is irreducible over F, has a real root, and has only constructible roots. Hence $P(X, mX)$ is the minimal polynomial over F of some constructible number α. By the results of section 5, the degree of $P(X, mX)$ is a power of 2.

To conclude that $P(X, MX)$ has degree a power of 2, write

$$P(X, MX) = \sum S_i(M) X^i$$

and note that each $S_i(M)$ has only finitely many distinct zeros. Hence for infinitely many of our choices of m, $S_i(m) \neq 0$ and $P(X, MX)$ has the same degree as $P(X, mX)$.

To complete the proof of the Theorem, let m be such that $Y = mX$ intersects $P(X, Y) = 0$ at some constructible point $(\alpha, \beta), \alpha \notin F$[27], i.e. $\beta = m\alpha$ and $P(\alpha, m\alpha) = 0$. Then there is a chain

---

[27] α is not in F whenever $P(X, mX)$ is irreducible.

$$F = F_0 \subseteq F_1 \ldots \subseteq F_n,$$

each $F_{i+1} = F_i[\sqrt{\alpha_i}]$ with $\alpha_i \in F_i$, and $\alpha \in F_n$. Choosing a chain of minimal length, we have $\alpha = a + b\sqrt{\alpha_{n-1}}$ for some $a, b \in F_{n-1}$. If we define $\overline{\alpha} = a - b\sqrt{\alpha_{n-1}}$, we see, as in section 5, $P(\overline{\alpha}, m\overline{\alpha}) = 0$, whence the roots occur in pairs. $\qquad\Box$

Let us consider a few examples. The first demonstrates why we needed to assume that the vertex of the pencil of lines having constructible intersections with the curve did not lie on the curve.

6.10 Example. $P(X, Y) = Y - X^3$.

Here $P(0,0) = 0$. $P(X, MX) = MX - X^3 = X(M - X^2)$ is reducible and, for each $m > 0$, $Y = mX$ intersects the curve at $X = 0, \pm\sqrt{m}$ (with $Y = 0, \pm(\sqrt{m})^3$, respectively), all constructible points. The degree of the curve is not a power of 2 and the points of intersection do not occur in pairs.

6.11 Example. $P(X, Y) = Y - X^3 - 1$.

$P(0,0) \neq 0$ and Theorem 6.9 applies. Since P is irreducible of degree not a power of 2, we know that some line passing through the origin intersects the curve in a nonconstructible point. This latter could also be seen by noting that every line passing through the origin of great enough slope[28] intercepts the graph of P in three distinct points.

6.12 Example. $P(X, Y) = X^4 + Y^4 - 1$.

Again $P(X, Y)$ is irreducible and $P(0,0) \neq 0$. Looking at $P(X, MX) = X^4 + M^4 X^4 - 1 = (1 + M^4)X^4 - 1$, we see that, for each m, $Y = mX$ intersects the curve at $X = \pm 1/\sqrt[4]{1 + m^4}$, which are constructible. $P(X, Y)$ has degree 4 and every line through the origin intersects P in exactly two points.

If we stick with $P(X, Y) = X^4 + Y^4 - 1$, but replace the origin by $(0, -2)$ as the vertex of our pencil of lines, P will still have degree 4 and, but for two tangent lines, any line passing through $(0, -2)$ that intersects the curve has two points of intersection. Hence the conclusion of Theorem 6.9 is satisfied. The hypothesis, however, is not. Consider the lines $Y = mX - 2$. At a point (x, y) of intersection with $P(X, Y) = 0$, x satisfies $P(X, mX - 2) = 0$. But

$$\begin{aligned}
P(X, mX - 2) &= X^4 + (mX - 2)^4 - 1 \\
&= X^4 + m^4 X^4 - 8m^3 X^3 + 24m^2 X^2 - 32mX + 15 \\
&= (1 + M^4 X^4) - 8m^3 X^3 + 24m^2 X^2 - 32mX + 15.
\end{aligned}$$

---

[28] $Y = mX$ has three intersections with the graph of P for $m > 3/\sqrt[3]{4}$. When m equals this value, $Y = mX$ is tangent to the curve at a point and crosses it elsewhere, thus intersecting the curve at two points. And for m less than this value, there is only one point of intersection. One method of determining this critical value of m, by the way, is to consider the equation $mX = X^3 + 1$ determining the point(s) of intersection and then looking where the discriminant of this cubic (Cf. the chapter on the cubic equation.) is positive, negative, or zero.

For $m = 2$, this yields

$$17X^4 - 64X^3 + 96X^2 - 64X + 15 = 0,$$

with 1 as a rational root, whence this equation factors into

$$(17X^3 - 47X^2 + 49X - 15)(X - 1) = 0.$$

The cubic factor has no rational, hence no constructible root. Thus the line $Y = 2X - 2$ intersects the curve $P(X,Y) = 0$ in the rational point $(1,0)$ and in a nonconstructible point.

Petersen follows his proof of Theorem 6.9 with the quick remark that, should the hypotheses of the Theorem hold for an arbitrary pencil of lines, then the degree of P is, in fact, 2. My first impression on reading this was that this result was an immediate corollary to Theorem 6.9. Failing that, the casualness of his announcement suggests it can be proven using no new ideas. For example, one has the following:

6.13 Theorem. *Let $P(X,Y)$ have coefficients in a finitely generated constructible number field F, P irreducible over F. Suppose the point(s) of intersection of the curve $P(X,Y) = 0$ with every constructible line is (are) constructible. Then the degrees of P in the variables X, Y are powers of 2.*

Proof. We consider only the case of the variable X.

For $b \in F$, define $P_b(X) = P(X,b)$. For all but finitely many values of b, the degree of $P_b(X)$ is the same as the degree of $P(X,Y)$ in the variable X.

So long as $P(X,Y)$ is not a horizontal line, there will be some interval (α, β) such that $P(X,Y)$ has an intersection with the horizontal lines $Y = b$ for all $b \in (\alpha, \beta)$. By Hilbert's Irreducibility Theorem, the set of $b \in (\alpha, \beta)$ with $b \in F$ for which $P_b(X)$ is irreducible is dense. Each such $P_b(X)$ has a constructible root and is the minimal polynomial thereof. Hence $P_b(X)$ has degree a power of 2. Hence the degree of $P(X,Y)$ in X is a power of 2.   □

This is nice. It allows us to rule out some curves of degree a power of 2 from having only constructible intersections with constructible lines, e.g. $X^8 + Y^6 - 1 = 0$ or $XY^3 - 1 = 0$. But it is still a far cry from Theorem 3.2. It does not handle the polynomial $X^4 + Y^4 - 1 = 0$ of Example 6.12 or the curves $X^2Y^2 - X \pm 1 = 0$, which satisfy the conclusions of Theorems 6.9 and 6.13.

I confess that I don't see how to prove a variant of Theorem 3.2 without using some new idea. If Petersen does so, it must be that the result is a corollary to his proof of Theorem 6.9 and not to the Theorem itself. I may have been too hasty in replacing his proof, which I have not yet mastered, by the quick reduction to Hilbert's later (1892) Irreducibility Theorem. But all is not lost. In a letter to Petersen dated 17 December 1870[29], Ludvig Sylow states

> I can now prove a theorem that contains something more, namely the following: the equation determining the intersections of a non-compound algebraic curve of nth degree with a straight line is in the same class as the general equations of the nth degree as far as its solution is concerned, as long as both parameters in the equation of the straight line are considered as independent variables. Thus, if the curve is of degree higher than 2, the equation cannot be solved by square roots... In a similar way one can more generally determine the system of conjugate substitutions belonging to the equation. Applying this procedure to the equation at hand, I find that the system must contain all possible substitutions... Thus I *think* (I cannot express myself with more certainty since I have not gone through the argument more than once) that the equation that determines the intersections with the lines in a pencil of lines is again in the same class as the general equations of nth degree as long as the vertex does not lie on the curve...

So Sylow says in effect that Galois theory provides such a new idea: if $P(X, Y)$ is as in the hypotheses of Theorem 6.13, then the Galois group of $P(X, MX + B)$ is the full symmetric group on the roots of P and thus has order $k!$, where P has k roots. However, the order of the Galois group is the order of the splitting field of P over F which is a power of 2 and such is a factorial only for $k = 1, 2$. For $k = 1$, $P(X, Y) = 0$ is a straight line, and for $k = 2$ the curve is a conic section.

Galois theory lies beyond the scope of the typical undergraduate History of Mathematics course and so must lie beyond the scope of the present book. So we shall settle here for the small taste of a proof of Theorem 3.2 we have thus far given, and the additional remark that Petersen's own proof was more elementary. The reader who, like me, intends to study Petersen's proof is referred to his book on the theory of equations. His dissertation has been published, but it is harder to find than the aforementioned later textbook. It might be added, however, that the two expositions differ in some points in the proof. This is discussed in Jesper Lützen's notes to the correspondence between Petersen and Sylow cited in footnote 29.

---

[29] Jesper Lützen, " The mathematical correspondence between Julius Petersen and Ludvig Sylow", in: Sergei Demidov, Menso Folkerts, David Rowe, and Christoph Scriba, eds., *Amphora: Festschrift für Hans Wussing zu seinem 65. Geburtstag*, Birkhäuser Verlag, Basel, 1992.

7 Concluding Remarks

In a later chapter we will discuss the solution of the cubic equation. Both it and the quartic equation can be solved by radicals, as known since the 16th century. No amount of work, however, would yield the solvability of the fifth degree equation by means of radicals and in 1813 Paolo Ruffini published an attempted proof that no such solution could be given. A little over a decade later, in 1826, Niels Henrik Abel published a correct but poorly expounded proof of this result. To quote S.G. Shanker[30]:

> And herein lies the key to Wantzel's failure to make any impact on the history of mathematics. Had Wantzel's proof been instrumental in bringing this new algebraic framework to the attention of his peers, his proof would undoubtedly have received considerable attention. But, of course, this was far from being the case. Rather, Wantzel's proof was formulated within the parameters of the existing work of Ruffini and Abel. It was thus an immediate consequence of the great break-throughs that had been achieved twenty years before. Indeed, Wantzel himself seems to have regarded his proof as little more than an off-shoot of the real issue at stake: the proof that the general equation of degree $n > 4$ cannot be solved algebraically. Certainly there is no evidence to suggest that he had been led into the field of modern algebra by his desire to solve the trisection problem. It was only his efforts to improve on Ruffini's proof that the general quintic equation is unsolvable by radicals which eventually led Wantzel to apply the same methods in his demonstration of the impossibility of the trisection problem.
>
> Wantzel's proof may have been the final step in the search to provide an adequate explanation for the impossibility of the trisection problem— thereby closing one of the most prolonged episodes in the history of mathematics— but it had only come about as a result of the developments in the theory of equations.

To this one should add that Wantzel published an improved version of Ruffini's impossibility proof in 1845. Galois's paper was finally published in 1846 and the first exposition, by Camille Jordan, of Galois theory was published in 1870. By 1895, the results were relegated to the list of applications of Galois theory, as mentioned in Klein's lectures and evidenced, as earlier stated, by Pierpont's neglecting to mention the trisection and duplication problems in his paper.

Something similar can be said of Petersen and his results. Unlike Wantzel, Petersen's primary interest was in the geometric constructibility problem. He had tried for years to trisect the angle before he finally set about proving the

---

[30] S.G. Shanker, "Wittgenstein's remarks on the significance of Gödel's Theorem", in: S.G. Shanker, ed., *Gödel's Theorem in Focus*, Croom Helm, London, 1988.

impossibility. His proof of Theorem 3.2 was almost complete by the time he learned of Galois's work from Ludvig Sylow in 1870.[31] The final difficulty in the proof for him was to show that if an equation of degree 2^n could be solved by square roots, then the roots could be expressed in terms of n different square roots (which could appear several times in the expression). As Sylow explained to Petersen,

> Galois has not written any paper especially about equations of degree 2^n, but he has written about certain equations of degree p^n, where p is an arbitrary prime... However, you will not find anything you can apply directly. The paper contains a general theory of algebraic equations, and especially of those that can be solved by radicals, but there is nothing especially devoted to equations that can be solved by square roots alone.

Petersen would learn Galois theory and include a couple of short chapters on the subject in his textbook on the theory of equations published in 1877. His textbooks were translated fairly quickly into several languages and attained a degree of popularity.

Admittedly, my personal library is weak on Galois theory and the theory of equations. Wantzel's results appear in some algebra texts, occasionally before the presentation of Galois theory (as in Herstein's *Topics in Algebra*[32]) and occasionally as an application of Galois theory (e.g. in Lisl Gaal's *Classical Galois Theory*[33] or van der Wærden's oft-cited *Moderne Algebra*). Wantzel's name is not mentioned in any of these works, but he is nowadays getting credited in the history books. Petersen's results have not been as well received. Perhaps it is my ignorance, but I know of no textbook other than Petersen's in which, e.g., Theorem 3.2 is proven. It is mentioned and Petersen's textbook on the theory of equations cited as a reference in Leonard Dickson's *Elementary Theory of Equations*[34], but no proof is offered.

One can see, both in the necessity of introducing deeper algebra in section 5 and in the historical fact that Petersen, on trying to solve the geometric problem, was being led to algebraic problems of the sort considered by Ruffini, Abel, and Galois that, had things been otherwise, these results could have had a major impact on mathematics, and field theory could have grown out of them. But this was not the case. Shanker is right. The impossibility proofs were regarded as more-or-less quick applications of powerful machinery already in place. However historically important the problems, their solutions have been shunted into a backwater of mathematics, kept alive mainly by the pædagogical instincts of men like Klein and Laugwitz who put a premium on elementary proofs.

---

[31] Cf. footnote 29.

[32] Blaisdell Publishing Company, New York, 1964.

[33] Chelsea Publishing Company, New York, 1971.

[34] John Wiley and Sons, New York, 1914.

A Chinese Problem

Word problems from old texts don't particularly interest me. The textbook authors who include problems from the Rhind Papyrus, al-Khwarezmi, or Fibonacci in their exercises would undoubtedly be horrified to read that I ignore these and leave it to the students to discover such on their own should they be interested. One problem, however, did catch my eye. In his textbook, *The History of Mathematics, An Introduction,* David Burton cites a problem from the *Mathematical Treatise in Nine Sections* (written 1247) of Ch'in Chushao (*c.* 1202 - 1261):

> There is a circular walled city of unknown diameter with four gates. A tree lies 3 *li* north of the northern gate. If one walks 9 *li* eastward from the southern gate, the tree becomes just visible. Find the diameter of the city.

The intriguing thing about this is that Burton goes on to say that, letting X^2 be the diameter of the city, Ch'in produces the tenth degree equation

$$X^{10} + 15X^8 + 72X^6 - 864X^4 - 11664X^2 - 34992 = 0. \qquad (1)$$

How did Ch'in come up with such an equation?

Of course, the equation is not really of degree 10 as one is interested in solving for X^2, not for X. The equation of actual interest is the fifth degree one,

$$X^5 + 15X^4 + 72X^3 - 864X^2 - 11664X - 34992 = 0. \qquad (2)$$

The question is: where did this equation come from?

Geometrically, the problem is easy to represent—cf. *Figure 1,* below. The condition that the tree at point C be just visible at B means that the line BC is tangent to the circle at some point which I have labelled D. For convenience, add a point E at the centre of the circle and the line segments DE and BE as in *Figure 2,* below.

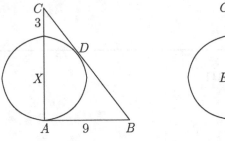
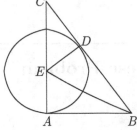

Figure 1   Figure 2

ABE and DBE are right triangles sharing the side BE. Moreover, their sides AE and DE being radii are equal. By the Pythagorean Theorem their third sides are also equal. Thus:

$$AB = BD = 9 \text{ and } AE = DE = \frac{X}{2}.$$

Now the triangle DEC is also a right triangle with two known sides:

$$DE = \frac{X}{2} \text{ and } CE = \frac{X}{2} + 3.$$

Thus the Pythagorean Theorem can again be applied:

$$CD^2 = CE^2 - DE^2 = \left(\frac{X}{2}+3\right)^2 - \left(\frac{X}{2}\right)^2 = 3X + 9.$$

Thus

$$BC = BD + DC = 9 + \sqrt{3X + 9}$$

and we can apply the Pythagorean Theorem once more to ABC to obtain

$$AC^2 = BC^2 - AB^2$$
$$= \left(9 + \sqrt{3X+9}\right)^2 - 9^2$$
$$= \left(81 + 18\sqrt{3X+9} + 3X + 9\right) - 81$$
$$= 3X + 9 + 18\sqrt{3X+9}. \tag{3}$$

But we also know

$$AC^2 = (X+3)^2 = X^2 + 6X + 9,$$

which with (3) yields

$$X^2 + 6X + 9 = 3X + 9 + 18\sqrt{3X+9},$$

whence

$$X^2 + 3X = 18\sqrt{3X + 9}.$$

Squaring both sides of this last yields

$$X^4 + 6X^3 + 9X^2 = 972X + 2916$$

and finally

$$X^4 + 6X^3 + 9X^2 - 972X - 2916 = 0. \tag{4}$$

Equation(4) is almost as ugly as Ch'in's equation (2), but it is not the same. However, we might as well use it to solve the problem. Graphing it on a calculator reveals it to have two roots, which we may determine graphically to be -3 and 9. (Or, in this case, we can test the divisors of 2916 to find the rational roots.) The positive solution 9 is the one that makes sense: the city has a diameter of 9 *li*.

Note that (4) factors into

$$(X - 9)(X + 3)(X^2 + 12X + 108) = 0, \tag{5}$$

the quadratic factor being irreducible. Ch'in's equation (2) has two solutions -6 and 9, and the factorisation

$$(X - 9)(X + 6)^2(X^2 + 12X + 108) = 0. \tag{6}$$

Despite the close relation between (5) and (6), I still did not see how Ch'in derived his equation, so I decided to look it up. First I checked Joseph Needham's *Science and Civilization in China,* in which Ch'in's name is given as Chhin Chiu-Shao and we learn that the Chinese name of the *Mathematical Treatise in Nine Sections* is *Shu Shu Chiu Chang.* Figure 55 on page 43 of volume III of Needham's magnum opus clearly belongs to the problem, but the problem itself is not to be found.

Ho Peng-Yoke's article on Ch'in Chiu-shao in the *Dictionary of Scientific Biography* is more informative. According to him, the *Shu-shu chiu-chang* covered more than twenty problems that "involve the setting up of numerical equations" and includes (1) in a short list of examples. With reference to this setting up of equations, Ho asserts that "sometimes Ch'in made his process unusually complicated" and cites our circular city problem as an example:

> Given a circular walled city of unknown diameter with four gates, one at each of the four cardinal points. A tree lies three *li* north of the northern gate. If one turns and walks eastward for nine *li* immediately leaving the southern gate, the tree becomes just visible. Find the circumference and the diameter of the city wall.

The language is more expansive than Burton's rendering, and there is the extra demand that one also find the circumference, but it is clearly the same problem. Replacing the 3 *li* north of the city by c and the 9 eastward by b, we read that Ch'in obtained the following equation for the square root of the diameter:

$$Y^{10} + 5cY^8 + 8c^2Y^6 - 4c(b^2 - c^2)Y^4 - 16c^2b^2Y^2 - 16c^3b^3 = 0. \quad (7)$$

There is a typographical error here as the last term yields -314928 instead of -34992 when one replaces b, c by their values $3, 9$, respectively. The term should read $-16c^3b^2$. There is no explanation how Ch'in obtained (7).

The next book I consulted was Yoshio Mikami's *The Develoment of Mathematics in China and Japan*. In the final paragraph of his chapter on Ch'in Chiu-shao, Mikami explained why no one tells us how Ch'in came up with (1) or (7):

> Notwithstanding that Ch'in Chiu-shao was very minute on the one hand in explaining the process of evaluation or root extracting of numerical equations, on the other he utterly neglected the description of the way to construct such equations by algebraical considerations from the given data in the problems.

This doesn't outright contradict Ho Peng-Yoke's remark that Ch'in's problems "involve the setting up of numerical equations", but it suggests caution in interpreting general remarks unaccompanied by source data.

As to the problem at hand, Mikami cites the following from book 8 of the *Su-shu Chiu-chang* or *Nine Sections of Mathematics*:

> There is a circular castle, whose circumference and diameter are unknown; it is provided with four gates and three miles (a) out of the north gate there is a large tree, which is visible from a point 9 miles (b) east of the south gate. What will be the lengths of the circumference and the diameter of the castle? Here the old value of π is to be used.

He further tells us that Ch'in offered up the equation,

$$X^{10} + 7cX^8 + 8c^2X^6 - 4(b^2 - c^2)X^2 - 2b^28c^2X^2 - 2b^28c^2b = 0,$$

where I have replaced a by c for readier comparison with equation (7). There are three typographical errors. The 7 in the second term of the equation should be a 5, a factor of c is missing from the fourth term, and, of course, the X^2 of the fourth term should be X^4. We are still not informed how Ch'in arrived at this equation, but at least we know now not to expect this.

The wording of the problem differs again from that given by Burton. The walled city has been replaced by a castle, the unit of distance replaced by the mile, and a value of π has been specified for the determination of the circumference. Nonetheless, this is recognisably the same problem. One assumes "castle" and "walled city" are expressed by the same word in Chinese and the correct translation would be determined by the context. Undoubtedly Mikami chose to replace the *li* by the mile because he was writing for a European audience and thought a more familiar unit of distance more appropriate. This choice and the ultimate diameter of 9 miles might suggest "walled city" as a more realistic rendering of the original word than "castle", but on the other hand, considering the size of some of the walls in China, one might want to stick with the word "castle". As for the "old value of π", it is a curious thing, but Ch'in uses no fewer than three different standard estimates of π:

the old value   3

Chang Hêng's value $\sqrt{10}$

the accurate value $\frac{22}{7}$.

Presumably he used the old value in this problem to make everything come out integral.

One more trip to the bookshelf produces Lǐ and Dù's *Chinese Mathematics, A Concise History.* This more modern book uses Pinyin instead of the older English Wade-Giles transliteration. Thus, instead of some variant spelling of "Ch'in Chiu-shao" we find "Qín Jiǔsháo" and his *Shùshū jiǔzhāng.* From Lǐ and Dù we learn some interesting things. First, the section from which our problem comes concerns military matters. Similar problems from Qín's contemporary Lǐ Yě concern forts, whence Mikami's "castle" may be a more accurate translation than might first have been apparent. Lǐ and Dù do not mention the problem, however, except in placing (1) among a list of higher order equations to be found among the 81 problems in Qín's book. The *Mathematical Treatise in Nine Sections* contains, alas, so much more important material to be discussed than a fairly easy word problem concerning circles and right triangles. This work contains the classical treatment of the Chinese Remainder Theorem, as well as what would be rediscovered centuries later in Europe as Horner's Method for approximating the solutions to higher degree polynomials[1]. The interest in our little problem is primarily taken to be that its tenth degree equation is the only one in the book of so high a degree. The problem is a mere novelty, paling in comparison with the rest, and is easily ignored in this context. A better context is the work of Lǐ Yě.

In switching from Qín to Lǐ, Lǐ and Dù implicitly repeat Mikami's warning not to expect to find Qín's derivation of (1) from the statement of the problem when they say that, "Although Qín Jiǔshāo's book *Mathematical Treatise in Nine Sections* has a systematic description of the 'method of extracting roots by iterated multiplication' it lacks a systematic procedure for writing down the equation". However, they do tell us that "the mathematicians of the eleventh to the thirteenth centuries in ancient China not only invented the 'method of extracting roots by iterated multiplication', which is a general method for solving higher degree equations, they also invented the general method for obtaining equations from given conditions. This was given the name 'technique of the celestial element'". They go on to say that Lǐ Yě was the first to give a systematic treatment of this technique in his *Sea Mirror of Circle Measurement* in 1248.

The Western reader who readily accepts Jean, John, and Johann Bernoulli as one man and yet might be amused by Ch'in, Chhin, and Qín will want to put his foot down on Lǐ. Mikami calls him Li Yeh, noting that one sometimes

---

[1] But see the chapter on Horner's method for some caveats to this latter assertion.

finds him listed as Li-Yeh Jin-king, Li being the family name, Yeh his personal name, and Jin-king his "familiar" name. In the *Dictionary of Scientific Biography,* he is listed as "Li Chih, also called Li Yeh". Jen-ch'ing is his "literary" name, Ching-chai the "appellation". The article, by Ho Peng-Yoke, explains further that Li's name was originally Li Chih, but that he changed it to Li Yeh so as not to have the same name as an earlier emperor. According to Lǐ and Dù, this change was accomplished through the elimination of a single stroke from the second character of his name. They use Pinyin and report that he thus changed his name from Lǐ Zhì to Lǐ Yě and that his literary name was Jingzhāi.

Although Lǐ and Qín were contemporaries and their works on equations complemented each other nicely, it is doubtful they ever met. They were separated politically as well as geographically, Lǐ living in the Mongol-occupied north and Qín in the Chinese south. Ho Peng Yoke adds that the two never mentioned one another and it is likely the two never even heard of each other. Thus it is striking that only a year after Qín's book appeared, problem 2 of chapter 7 of Lǐ's *Sea Mirror of Circle Measurement* reads (according to Lǐ and Dù):

> [Assume there is a circular fort of unknown diameter and circumference,] person *A* walks out of the south gate 135 steps and person *B* walks out of the east gate 16 steps and then they see one another. [What is the diameter?]

The diagram one obtains from this is not quite the same as that for Qín's problem, but it contains the same elements— a circle and a right triangle the hypotenuse of which is tangent to the circle. The *Dictionary of Scientific Biography* cites several further examples, including the following:

> *A* leaves the western gate [of a circular city wall] and walks south for 480 *pu*. *B* leaves the eastern gate and walks straight ahead a distance of 16 *pu*, when he just begins to see *A*. Find [the diameter of the city wall] as before.

Graphically, this is the same problem as Qín's.

Both Ho Peng-Yoke and Lǐ and Dù give an example of Lǐ's deriving an equation for one of his word problems of this form. Lǐ and Dù note that Lǐ gives five solutions to the first problem cited, and they repeat the second of these, which solution I am quite taken with. Therefore, I shall apply it to a generic version of Qín's problem, as given in *Figure 3*. Again, we let X denote the diameter of the circle. Our earlier solution used the Pythagorean Theorem a couple of times. Lǐ calculates the area of

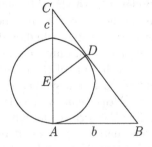

Figure 3

the triangle ABC twice. On the one hand, it is half the area of the rectangle determined by AB and AC,

$$\Delta = \frac{1}{2}AB \times AC = \frac{1}{2}b(X+c)$$

But it is also equal to the sum of the areas of the triangles ABE and BCE:

$$\Delta = \frac{1}{2}b\frac{X}{2} + \frac{1}{2}BC \times DE$$

$$= \frac{bX}{4} + BC \times \frac{X}{4}$$

$$= \frac{bX}{4} + \sqrt{b^2 + (X+c)^2}\,\frac{X}{4},$$

using the Pythagorean Theorem. (This step is easier with the problem cited by Lǐ and Dù.) Equating the two expressions,

$$\frac{b}{2}(X+c) = \frac{bX}{4} + \frac{X}{4}\sqrt{b^2 + (X+c)^2}$$

$$\frac{bX}{4} + \frac{bc}{2} = \frac{X}{4}\sqrt{b^2 + (X+c)^2}$$

$$bX + 2bc = X\sqrt{b^2 + (X+c)^2}$$

$$b^2X^2 + 4b^2Xc + 4b^2c^2 = X^2b^2 + X^2(X+c)^2$$

$$4b^2c(X+c) = X^2(X+c)^2 \tag{8}$$

$$4b^2c = X^3 + cX^2 \tag{9}$$

i.e.

$$X^3 + cX^2 - 4b^2c = 0. \tag{10}$$

If one chooses not to cancel $X+c$ in the passage from (8) to (9), then (10) will be replaced by the equation

$$X^4 + 2cX^3 + c^2X^2 - 4cb^2X - 4c^2b^2 = 0. \tag{11}$$

For $c=3, b=9$, equation(11) becomes equation (4).

In our derivations of (4) and (11) we have made two applications of the Pythagorean Theorem and one application of the Pythagorean Theorem in combination with two area computations. We can also solve the problem by pairing an application of the Pythagorean Theorem with the observation that the triangles ABC and DEC are similar:

$$CD = \sqrt{\left(\frac{X}{2}+c\right)^2 - \left(\frac{X}{2}\right)^2}$$

$$= \sqrt{\left(\frac{X}{2}\right)^2 + 2\frac{X}{2}c + c^2 - \left(\frac{X}{2}\right)^2}$$

$$= \sqrt{c^2 + cX}.$$

But
$$\frac{CD}{DE} = \frac{CA}{AB},$$

i.e.
$$\frac{\sqrt{c^2 + cX}}{X/2} = \frac{X + c}{b}$$
$$\sqrt{c^2 + cX} = \frac{X(X + c)}{2b}$$
$$c^2 + cX = \frac{X^2(X + c)^2}{4b^2},$$

which is equivalent to (8) above and thus yields (10). The ratio
$$\frac{CD}{CE} = \frac{CA}{BC}$$

also yields (10).

For $c = 3, b = 9$, (10) becomes
$$X^3 + 3X^2 - 972 = 0. \tag{12}$$

Our penultimate solution: We can avoid appeal to the Pythagorean Theorem by making even stronger use of similar triangles as follows.
$$\frac{AC}{CD} = \frac{AB}{DE} = \frac{CB}{CE}. \tag{13}$$

The first of these equations reads
$$\frac{X + c}{CD} = \frac{b}{X/2},$$

whence
$$CD = \frac{X}{2}(X + c)\frac{1}{b} = \frac{X(X + c)}{2b}. \tag{14}$$

The second equation of (13) reads
$$\frac{b}{X/2} = \frac{b + CD}{X/2 + c}.$$

Plugging the value of CD from (14) into this yields

$$\frac{2b}{X} = \frac{b + X(X + c)/(2b)}{X/2 + c}$$

$$2b\left(\frac{X}{2} + c\right) = bX + \frac{X^2(X + c)}{2b}$$

$$b^2(X + 2c) = b^2 X + \frac{X^2}{2}(X + c)$$

$$b^2 X + 2b^2 c = b^2 X + \frac{X^2}{2}(X + c)$$

$$2b^2 c = \frac{X^2}{2}(X + c)$$

$$4b^2 c = X^3 + cX^2$$

$$X^3 + cX^2 - 4b^2 c = 0,$$

which is just equation (10).

This may well be the simplest solution. However my question was not to solve Ch'in's problem but to see how on earth he arrived at the complicated equation (2). Toward this end, let me draw a new diagram and present one last solution to the problem. Extend the triangle as as in *Figure 4*.

Then FG is tangent to the circle and perpendicular to GAB. Thus $GA = \frac{X}{2}$.

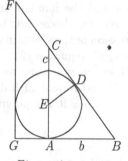

Figure 4

Applying similarity,

$$\frac{FG}{CA} = \frac{BG}{BA},$$

i.e.

$$\frac{FG}{X + c} = \frac{b + X/2}{b} = \frac{2b + X}{2b},$$

whence

$$FG = \frac{(X + c)(2b + X)}{2b}. \tag{15}$$

But also

$$\frac{FB}{CE} = \frac{GB}{DE},$$

i.e.

$$\frac{FB}{X/2 + c} = \frac{X/2 + b}{X/2},$$

whence

$$FB = \frac{2c + X}{2} \frac{2b + X}{X}. \tag{16}$$

Now we return to applying the Pythagorean Theorem,

$$FB^2 = FG^2 + GB^2,$$

i.e.

$$\frac{(2c + X)^2(2b + X)^2}{4X^2} = \frac{(X + c)^2(2b + X)^2}{4b^2} + \frac{(2b + X)^2}{4}$$
$$b^2(2c + X)^2(2b + X)^2 = X^2(X + c)^2(2b + X)^2 + b^2X^2(2b + X)^2$$
$$(4b^2c^2 + 4b^2cX + b^2X^2)(2b + X)^2 = (X^4 + 2cX^3 + c^2X^2 + b^2X^2)(2b + X)^2,$$

and we have

$$(X^4 + 2cX^3 + c^2X^2 - 4b^2cX - 4b^2c^2)(2b + X)^2 = 0. \tag{17}$$

Now, if we cancel $(2b + X)^2$, which is zero only at the uninteresting negative value $-2b$, we once again obtain equation (11) whose polynomial factors into $X + c$ and the familiar cubic of (10) by applying factoring by grouping. If, however, we choose not to cancel $(2b + X)^2$ from (17) but to multiply the expression out, we obtain a sixth degree equation which is even more complex than Qín's equation (2).

If we do not want to outdo Qín, we can retain only one of the factors $X + 2b$ and have a fifth degree equation. Alas, it is still not Qín's equation. For $c = 3, b = 9$, the polynomial (17) has the factorisation

$$(X - 9)(X + 18)^2(X + 3)(X^2 + 12X + 108),$$

which will not yield Qín's polynomial on removing any linear factor.

My search ended in failure. In three different sources I found not merely three translations of Qín's problem, but three different statements of the problem: two asked for the circumference as well as the diameter, one gave the value of π to be used, and one was given in miles. The two sources giving Qín's general solution gave different solutions— both incorrect, due to non-overlapping typographical errors. At this stage, only the very worst students would not be longing for the primary reference, which, apparently, has yet to be translated into English. An internet search had revealed the existence of another book on Chinese mathematics that might prove interesting, but being offered at over $200 for a used copy, I decided it was out of my price range for so minor a point.[2] This was Ulrich Libbrecht, *Chinese Mathematics in the Thirteenth Century*, MIT Press, Cambridge (Mass.), 1973. In the months that passed since the above account was typeset and this book submitted, Libbrecht's book has been reprinted by Dover Publications, Inc., Mineola (New York), 2005. As soon as I learned of it I acquired a copy and am pleased to report that it has some very good information on this word problem, including a derivation of Qín's equation.

---

[2] I believe I tried to borrow a copy through interlibrary loan, but that succeeds only half the time, even for books and articles available across town.

Libbrecht's book bears the subtitle, *The Shu-shu chiu-chang of Ch'in Chiu-shao*, but is not a translation of this book. It translates some passages and treats a number of Ch'in's problems and solutions, but also embedds the whole in a general discussion of Chinese mathematics of the period, and includes a sub-monograph on the Chinese Remainder Theorem. It also includes a biography of Ch'in from which we learn that he was a remarkably evil man whose attempt to have his son murdered makes the bickering Bernoulli brothers look like poster boys for fraternal affection. We also learn that his "courtesy" name was Tao-ku. Recalling the familiar name, literary name, and appellation cited on page 138, I am beginning to wonder how many aliases a Chinese scholar was allowed to have.

Getting down to business, let me cite Libbrecht's statement of the problem from pp. 134 - 135 of his book:

> There is a round town of which we do not know the circumference and the diameter. There are four gates [in the wall]. Three *li* outside the northern [gate] there is a high tree. When we go outside the southern gate and turn east, we must cover 9 *li* before we can see the tree. Find the circumference and the diameter of the town ($\pi = 3$).[3]

On page 136 the same illustration of the problem as appears in Needham can be found. Libbrecht does not translate Ch'in's discussion of the problem, noting only that Ch'in says that if x^2 is the diameter then x satisfies the equation (7) (with the typo corrected) and that x is 3. After a brief discussion of Li Yeh's approach applied to this problem, he shows how Pai Shang-shu obtained Ch'in's equation in the 1960s. As in deriving (17), this is accomplished by redrawing *Figure 3*, this time by extending a horizontal line from E to a point F on the hypotenuse of the large triangle as in *Figure 5* on the right.

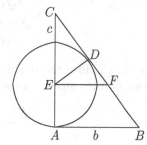

Figure 5

One now toys with the similarity of the triangles ABC, EFC, and DEC. The first pair yields

$$\frac{CE}{EF} = \frac{CD}{DE},$$

i.e.

$$\frac{X/2 + c}{EF} = \frac{\sqrt{(X/2 + c)^2 - (X/2)^2}}{X/2}, \qquad (18)$$

whence

---

[3] Bracketed insertions are Libbrecht's.

$$EF = \frac{(X/2)(X/2+c)}{\sqrt{Xc+c^2}}$$

$$= \frac{X(X+2c)}{4\sqrt{c(X+c)}}. \tag{19}$$

The second pair yields

$$\frac{CE}{EF} = \frac{CA}{AB},$$

i.e.

$$\frac{X/2+c}{EF} = \frac{X+c}{b}, \tag{20}$$

whence

$$EF = \frac{b(X/2+c)}{X+c}$$

$$= \frac{b(X+2c)}{2(X+c)}. \tag{21}$$

From (19) and (21) it follows

$$\frac{X(X+2c)}{4\sqrt{c(X+c)}} = \frac{b(X+2c)}{2(X+c)}$$

$$\frac{X^2(X+2c)^2}{16c(X+c)} = \frac{b^2(X+2c)^2}{4(X+c)^2}. \tag{22}$$

If we now reduce the denominators and conveniently overlook the fact that we can do the same with the numerators, this last becomes

$$\frac{X^2(X+2c)^2}{4c} = \frac{b^2(X+2c)^2}{X+c}. \tag{23}$$

Cross-multiplying yields

$$X^2(X+2c)^2(X+c) = 4b^2c(X+2c)^2,$$

whence

$$(X+2c)^2(X^3+cX^2-4b^2c) = 0. \tag{24}$$

Multiplying this out yields the corrected form of (7). [Alternatively, one can plug $b = 9, c = 3$ into (21) and factor the cubic to obtain the factored form (6) of Ch'in's equation.]

The derivation is not completely satisfying. If one had equated (18) and (20) or deleted the common numerator of (23), one would have obtained the cubic portion of (24) immediately. Pan Shang-shu says,

... was his intention to construct an equation of higher degree to set a record? If so, he should not have reduced the denominators in (22) and he would have gotten an equation of the twelfth degree[4]. Moreover, he reduces only the denominators and not the numerators, which is difficult to explain.

This sounds rather critical, but on deeper reflexion it is actually something of a complement. Compare this with Libbrecht's more positive sounding remark:

> We shall see that Ch'in Chiu-shao sometimes constructed equations of a degree higher than necessary for solving his problems; the only explanation is that he wanted to prove that he was able to solve them. And here we meet the true mathematician as opposed to the technologist.[5]

Libbecht states a few pages later (p. 16), regarding the practicality of Chinese mathematics:

> This necessarily practical attitude was an impediment to the unfolding of genius of some mathematicians; it is a striking fact that, as mentioned earlier, Ch'in Chiu-shao twists and turns to construct practical problems (which do not look practical at all) in order to get equations of a degree high enough to prove his ability in solving them. All this points to one of the main reasons for the final stagnation of Chinese mathematics. Indeed, it is, in the traditional Chinese mind, foolish to solve an equation of the tenth degree when there is no practical problem that requires it.

So there we have it. We do not know exactly how Qín derived his equation from the problem. We have one way of doing so, but after experiencing the various derivations of Lǐ Yě's equation, we know that there is no guarantee it was *the* derivation: there are other such derivations that work. We can be pretty sure, however, that the derivation followed these general lines and that should suffice. I consider myself lucky in that I used the "wrong" figure in trying to derive his equation (and with 20‑20 hindsight I see clearly that the factor $X + 6 = 2(X/2 + 3)$ in the factorisation (6) should have driven me to add the line EF of *Figure 5* right away), for had I been successful I would not have bothered looking up Needham, Mikami, Lǐ and Dù, Ho Peng-Yoke's articles, or, ultimately, Libbrecht.

We can draw some lessons about researching a term paper from my little oriental journey.

Even the best of secondary references can lead one astray. In Ho Peng-Yoke's article on Ch'in Chiu-shao in the *Dictionary of Scientific Biography,* after finding Ch'in's equation (7), one reads

> It is interesting to compare this with an equivalent but simpler expression given by Ch'in's contemporary Li Chih in the form

---

[4] Pan Shang-shu deals with the original equation in which the diameter is represented by X^2 instead of X.

[5] Libbrecht, p. 9. In a footnote he quotes Youschkevitch to the same effect.

$$X^3 + cX^2 - 4cb^2 = 0.$$

If one did not go to the bookshelf and pull down the volume containing Ho's article on Li Chih or consult Lǐ and Dù, one could easily read too much into this and conclude, for $b = 9, c = 3$, that Li Chih produced equation (12) for Ch'in's problem, when, in all likelihood, he never saw this latter.

The Cubic Equation

1 The Solution

Every good history of math book will present the solution to the cubic equation and tell of the events surrounding it. The book will also mention, usually without proof, that in the case of three distinct roots the solution must make a detour into the field of complex numbers. In these notes we prove this result and also discuss a few other nuances often missing from the history books.

To solve a general cubic equation,

$$X^3 + rX^2 + sX + t = 0, \tag{1}$$

one starts by applying the cubic analogue of completing the square, substituting $X = Y - \frac{r}{3}$:

$$
\begin{array}{llll}
\left(Y - \frac{r}{3}\right)^3 & = Y^3 - rY^2 + & \frac{r^2}{3}Y & - & \frac{r^3}{27} \\
r\left(Y - \frac{r}{3}\right)^2 = & rY^2 - & \frac{2r^2}{3}Y & + & \frac{r^3}{9} \\
s\left(Y - \frac{r}{3}\right) = & & sY & - & \frac{rs}{3} \\
t & = & & & t \\
\Sigma & = Y^3 + & \left(s - \frac{r^2}{3}\right)Y + \left(\frac{2r^3}{27} - \frac{rs}{3} + t\right). &
\end{array}
$$

Thus it suffices to solve equations of the form

$$Y^3 + pY + q = 0. \tag{2}$$

An inspired substitution does the rest. Let $Y = a - b$,

$$(a - b)^3 + p(a - b) + q = 0,$$

whence

$$a^3 - 3a^2b + 3ab^2 - b^3 + p(a - b) + q = 0,$$

i.e.

$$a^3 - b^3 + q - 3ab(a - b) + p(a - b) = 0.$$

If one now chooses a, b so that

$$a^3 - b^3 + q = 0 \quad \text{and} \quad 3ab = p, \tag{3}$$

one will have constructed the desired solution Y to (2) from which the solution X to (1) may be had.

For later reference, note that when $p \neq 0$ it follows from the second equation of (3) that $a \neq 0$ and

$$b = \frac{p}{3a} \tag{4}$$

Of more immediate interest is the first equation of (3) which yields, on multiplying by a^3,

$$a^6 - a^3 b^3 + a^3 q = 0,$$

i.e.

$$a^6 + qa^3 - \frac{p}{27} = 0,$$

which is quadratic in a^3. The quadratic formula yields

$$a^3 = \frac{-q \pm \sqrt{q^2 + \frac{4}{27}p^3}}{2}. \tag{5}$$

The first equation of (3) yields

$$b^3 = \frac{q \pm \sqrt{q^2 + \frac{4}{27}p^3}}{2}, \tag{6}$$

whence one has the solution

$$Y = a - b = \sqrt[3]{\frac{-q \pm \sqrt{q^2 + \frac{4}{27}p^3}}{2}} - \sqrt[3]{\frac{q \pm \sqrt{q^2 + \frac{4}{27}p^3}}{2}}. \tag{7}$$

If equation (2) has one real and two complex roots, formula (7) produces the real root. It may not be presented in the most transparent manner, but it is correct. If, however, equation (2) has three distinct real roots, then i) the *discriminant* $q^2 + \frac{4}{27}p^3$ is negative and the solution requires one to deal with complex numbers, and ii) formula (7), hastily applied, may not yield a real number at all.

The latter problem is easily explained. Complex numbers have three cube roots and one must choose matching pairs of roots of a^3 and b^3 in (5) and (6) in order that (7) actually yield a root. A better strategy is to find a cube root a of (5) and choose b according to (4).

An even better approach when the discriminant is negative is to apply (4) to show that, if $a - b$ is a solution, then b is the negative of the conjugate of a, whence

$$a - b = a - (-\bar{a}) = a + \bar{a} = 2Re\,(a). \tag{8}$$

To see that this is the case, note that

$$b = \frac{p}{3a} = \frac{p\bar{a}}{3|a|^2}. \tag{9}$$

Now, for $x = a - b$ to be real, we must have $Im\,(a) = Im\,(b)$. But then

$$Im\,(a) = Im\,(b) = \frac{p}{3|a|^2} Im\,(\bar{a}) = \frac{p}{3|a|^2}\,(-Im\,(a)),$$

whence $1 = -\frac{p}{3|a|^2}$ and $b = -\bar{a}$ by (9).

2 Examples

Let me illustrate the situation with a couple of simple examples.

2.1 Example. $X^3 - 3X^2 + 4X - 3 = 0$.

The substitution $X = Y + 1$ transforms this into

$$Y^3 + Y - 1 = 0.$$

Here $p = 1, q = -1$, whence the discriminant

$$q^2 + \frac{4}{27}p^3 = 1 + \frac{4}{27} = \frac{31}{27}$$

is positive. The equation will have one real and two complex solutions. Letting $Y = a - b$ and so on, we have

$$a^3 = \frac{1 \pm \sqrt{\frac{31}{27}}}{2} = \frac{1 \pm \frac{1}{9}\sqrt{93}}{2} = \frac{9 \pm \sqrt{93}}{18}.$$

By formula (7),

$$Y = \sqrt[3]{\frac{9 + \sqrt{93}}{18}} - \sqrt[3]{\frac{-9 + \sqrt{93}}{18}} = .6823278038,$$

which is indeed a real root of the equation in question.

2.2 Example. $Y^3 - 3Y - 1 = 0$.

Here $p = -3, q = -1$, whence the discriminant

$$q^2 + \frac{4}{27}p^3 = 1 + \frac{4}{27}(-27) = 1 - 4 = -3$$

is negative. The equation will have three distinct real solutions. Letting $Y = a - b$ and so on, we have

$$a^3 = \frac{1 + \sqrt{-3}}{2}, \quad b^3 = \frac{-1 + \sqrt{-3}}{2}.$$

If we apply (7) and use a calculator we get

$$a = .9396926208 + .3420201433i$$

$$b = .7660444431 + .6427876097i$$

and $a - b$ is neither real nor a solution to the equation. However, if one multiplies b by $\frac{-1+\sqrt{-3}}{2}$, which is a cube root of 1, one obtains another cube root of b^3 :

$$b_2 = b\frac{-1 + \sqrt{-3}}{2} = -.9396926208 + .3420201433i,$$

and

$$a - b_2 = 2Re(a) = 1.879385242$$

is a real solution to the equation. The other solutions are

$$2Re\left(a\frac{-1 + \sqrt{-3}}{2}\right) = -1.532088886$$

$$2Re\left(a\frac{-1 - \sqrt{-3}}{2}\right) = -.3472963553.$$

2.3 Example. $Y^3 - 3Y + 2 = 0$.

Here $p = -3, q = 2$, whence the discriminant

$$q^2 + \frac{4}{27}p^3 = 4 + \frac{4}{27}(-27) = 4 - 4 = 0$$

vanishes. The equation will have three real roots, one repeated. We have

$$a^3 = \frac{-2 + \sqrt{0}}{2} = -1, \quad b^3 = \frac{2 + \sqrt{0}}{2} = 1,$$

whence $a = -1, b = 1$ and $a - b = -1 - 1 = -2$ is the non-repeating root of the equation. The repeating root may be found either by dividing $Y^3 - 3Y + 2$ by $Y + 2$ and solving the quadratic, or by choosing for a one of the other cube roots of -1, e.g.

$$a = \frac{1 + \sqrt{-3}}{2}$$

and using (4) to define b:

$$b = \frac{p}{3a} = \frac{-3}{3a} = \frac{-1}{a} = -\frac{1 - \sqrt{-3}}{2} = \frac{-1 + \sqrt{-3}}{2}.$$

Then

$$a - b = \frac{1 + \sqrt{-3}}{2} - \frac{-1 + \sqrt{-3}}{2} = \frac{2}{2} = 1.$$

3 The Theorem on the Discriminant

3.1 Theorem. *Let the equation*

$$Y^3 + pY + q = 0 \tag{10}$$

be given, with p, q real numbers and define the discriminant

$$D = q^2 + \frac{4}{27}p^3.$$

i. if $D > 0$, the equation has one real and two distinct complex roots;
ii. if $D = 0$, the equation has three real roots, at least two of which are identical;
iii. if $D < 0$, the equation has three distinct roots.

We will actually prove the converses to these three implications. As the conditions on the discriminants are mutually exclusive this is sufficient.

Let α, β, γ be the (possibly complex) roots of (10). Then

$$Y^3 + pY + q = (Y - \alpha)(Y - \beta)(Y - \gamma)$$
$$= Y^3 - (\alpha + \beta + \gamma)Y^2 + (\alpha\beta + \beta\gamma + \gamma\alpha)Y - \alpha\beta\gamma. \tag{11}$$

3.2 Lemma. *The following hold:*
i. $\gamma = -(\alpha + \beta)$
ii. $p = -(\alpha^2 + \alpha\beta + \beta^2)$
iii. $q = \alpha\beta(\alpha + \beta)$.

Proof. i. Equating coefficients in (11) yields $\alpha + \beta + \gamma = 0$.
ii. Again,

$$p = \alpha\beta + \beta\gamma + \gamma\alpha, \text{ by (11)}$$
$$= \alpha\beta + (\alpha + \beta)\gamma$$
$$= \alpha\beta - (\alpha + \beta)^2, \text{ by i}$$
$$= \alpha\beta - \alpha^2 - 2\alpha\beta - \beta^2$$
$$= -\alpha^2 - \alpha\beta - \beta^2.$$

iii. Again,

$$q = -\alpha\beta\gamma, \text{ by (11)}$$
$$= -\alpha\beta\,(-\alpha - \beta)$$
$$= \alpha\beta\,(\alpha + \beta). \qquad \qquad \square$$

3.3 Lemma. $q^2 + \frac{4}{27}p^3 = \alpha^2\beta^2\,(\alpha+\beta)^2 - \frac{4}{27}\left(\alpha^2 + \alpha\beta + \beta^2\right)^3.$

Proof of the Theorem. The easiest case is when two of the roots are equal: $\alpha = \beta$. By Lemma 3.3,

$$D = \alpha^2\alpha^2\,(\alpha + \alpha)^2 - \frac{4}{27}\left(\alpha^2 + \alpha^2 + \alpha^2\right)^3$$

$$= \alpha^4\,(2\alpha)^2 - \frac{4}{27}\left(3\alpha^2\right)^3$$

$$= 4\alpha^6 - \frac{4}{27}27\alpha^6 = 0.$$

Digressing a moment, note that the solution given by the formulæ

$$y = a - b, \quad a^3 = \frac{-q + \sqrt{D}}{2}, \quad b = \frac{p}{3a}$$

is

$$y = \sqrt[3]{-\frac{\alpha\beta\,(\alpha+\beta)}{2}} - \frac{p}{3\sqrt[3]{-\frac{\alpha\beta(\alpha+\beta)}{2}}}$$

$$= \sqrt[3]{\frac{-\alpha^2\,(2\alpha)}{2}} - \frac{-3\alpha^2}{3\sqrt[3]{-\frac{\alpha^2\,(2\alpha)}{2}}}$$

$$= -\alpha - \frac{-\alpha^2}{-\alpha} = -\alpha - \alpha = -2\alpha = \gamma,$$

the (possibly) non-repeating root.

The case of two complex roots is also fairly elementary. Suppose α, β are non-real roots. Then $\beta = \overline{\alpha}$ is the conjugate of α and we have

$$D = \alpha^2\overline{\alpha}^2\,(\alpha + \overline{\alpha})^2 - \frac{4}{27}\left(\alpha^2 + \alpha\overline{\alpha} + \overline{\alpha}^2\right)^3$$

$$= |\alpha|^4\,(2Re\,(\alpha))^2 - \frac{4}{27}\left((\alpha + \overline{\alpha})^2 - \alpha\overline{\alpha}\right)^3$$

$$= 4|\alpha|^4 Re\,(\alpha)^2 - \frac{4}{27}\left((2Re\,(\alpha))^2 - |\alpha|^2\right)^3$$

$$= 4|\alpha|^4 Re\,(\alpha)^2 + \frac{4}{27}\left(|\alpha|^2 - 4Re\,(\alpha)^2\right)^3$$

$$> 4Re\,(\alpha)^4 Re\,(\alpha)^2 + \frac{4}{27}\left(Re\,(\alpha)^2 - 4Re\,(\alpha)^2\right)^3,$$

since $|\alpha| > |Re\,(\alpha)|$, whence

$$D > 4Re\,(\alpha)^6 + \frac{4}{27}\left(-3Re\,(\alpha)^2\right)^3$$
$$> 4Re\,(\alpha)^6 - \frac{4}{27}27Re\,(\alpha)^6$$
$$> 0.$$

Another digressive note: The classic solution, as already noted, produces $2Re\,(a)$ as the solution to (10). This being real, it produces γ. Now $\gamma = -\alpha - \beta = -\alpha - \overline{\alpha} = -2Re\,(\alpha)$, whence

$$Re\,(\alpha) = -Re\,(a).$$

This doesn't seem particularly important, but it is nice nonetheless.

The case of three distinct real roots it the trickiest one. We begin by defining a function of two real variables:

$$f(x,y) = x^2y^2\,(x+y)^2 - \frac{4}{27}\left(x^2 + xy + y^2\right)^3.$$

A glance at the three-dimensional graph of this function strongly suggests the truth of the following lemma.

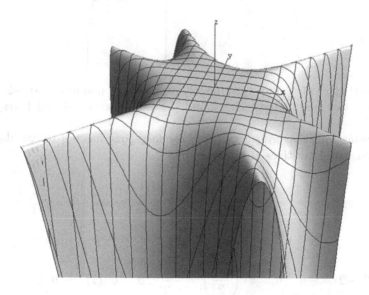

3.4 Lemma. *For all real numbers $\alpha, \beta, f(\alpha, \beta) \leq 0$.*

Proof. For $\alpha = 0$, we have $f(0, \beta) = -\frac{4}{27}\beta^6 \leq 0$, with equality only when $\beta = 0$.

For $\alpha \neq 0$ and any choice of β, we can write $\beta = k\alpha$ for some k. Then

$$f\left(\alpha, \beta\right) = f\left(\alpha, k\alpha\right) = \alpha^2 k^2 \alpha^2 \left(\alpha + k\alpha\right)^2 - \frac{4}{27} \left(\alpha^2 + k\alpha^2 + k^2\alpha^2\right)^3$$

$$= k^2 \left(1 + k\right)^2 \alpha^6 - \frac{4}{27} \left(1 + k + k^2\right)^3 \alpha^6$$

$$= [k^2 + 2k^3 + k^4 - \frac{4}{27} \left(1 + k + k^2\right)^3]\alpha^6$$

$$= g\left(k\right)\alpha^6, \tag{12}$$

where

$$g\left(x\right) = x^2 + 2x^3 + x^4 - \frac{4}{27}\left(1 + x + x^2\right)^3$$

$$= \frac{-4x^6 - 12x^5 + 3x^4 + 26x^3 + 3x^2 - 12x - 4}{27}. \tag{13}$$

On the pocket calculator the graph of $y = g\left(x\right)$ looks something like the following(but without the vertical tangents at the inflexion points necessitated by my lack of familiarity with graphics in LaTeX):

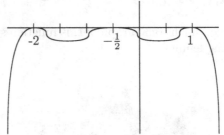

We see that $g\left(x\right) \le 0$ everywhere, there are three maxima— at $-2, -\frac{1}{2}$, and 1— and two minima between -2 and $-\frac{1}{2}$ and between $-\frac{1}{2}$ and 1, respectively.

To obtain a quick rigorous proof of these facts, one has but to calculate g and its first two derivatives at a few points:

$$g\left(-2\right) = g\left(\frac{-1}{2}\right) = g\left(1\right) = 0$$

$$g'\left(-2\right) = g'\left(\frac{-1}{2}\right) = g'\left(1\right) = 0$$

$$g''\left(-2\right) = -6 < 0, \quad g''\left(\frac{-1}{2}\right) = -\frac{3}{2} < 0, \quad g''\left(1\right) = -6 < 0.$$

Hence we indeed have local maxima with value 0 at $-2, -\frac{1}{2}$, and 1. On general principles, the other extrema must be minima and we have $g\left(x\right) \le 0$ for all x. In particular, $g\left(k\right)\alpha^6 \le 0$ and, by (12) $f\left(\alpha, \beta\right) \le 0$. [For the algebraically rigorous, I note that, after factoring $x + 2, x + \frac{1}{2}, x - 1$ from

$$g'(x) = \frac{-24x^5 - 60x^4 + 12x^3 + 78x^2 + 6x - 12}{27},$$

one is left with the quadratic polynomial

$$-\frac{12}{27}\left(2x^2 + 2x - 1\right),$$

which has the roots $\frac{-1\pm\sqrt{3}}{2}$ and these can readily be checked to yield minima.]   □

Alternate Proof that $g(x) \leq 0$. The use of Calculus to prove a simple algebraic inequality can be avoided. Because $-2, et\ al.$ are roots not only of g but also of g', these are double roots of g. This means that, upon factoring, g will assume the form

$$g(x) = K(x+2)^2\left(x+\frac{1}{2}\right)^2(x-1)^2 \tag{14}$$

for some constant K. But K can be determined by evaluating g at any point using (13) and (14). For example,

$$g(-1) = -\frac{4}{27}, \text{ by (13).}$$

But we also have

$$g(-1) = K(1)^2\left(-\frac{1}{2}\right)^2(-2)^2 = K,$$

whence $K = -\frac{4}{27}$ and

$$g(x) = -\frac{4}{27}(x+2)^2\left(x+\frac{1}{2}\right)^2(x-1)^2 \tag{15}$$

$$\leq 0,$$

with equality only at $x = -2, -\frac{1}{2}, 1$. Now, the appeal to the Calculus in establishing (15) can be bypassed via the simple expedient of dividing the numerator of (13) by $x+2, x+2, x+\frac{1}{2}, x+\frac{1}{2}, x-1$ and $x-1$ in succession.   □

Getting back to the proof of the Theorem, it follows from Lemma 3.4 that if (10) has three distinct real roots the discriminant is non-positive. To complete the proof, we must yet show that if the roots are distinct, the discriminant does not take on the value 0; in other words, if there are three real roots and the discriminant is 0, then two of the roots are identical.

Suppose α, β, γ are real roots of (2) and suppose further that $D = f(\alpha, \beta) = 0$.

If $\alpha = 0$, then $0 = f(0, \beta) = -\frac{4}{27}\beta^6$ and $\beta = 0$. Hence $\alpha = \beta$.

If $\alpha \neq 0$, we can write $\beta = k\alpha$. Now $f(\alpha, k\alpha) = 0$ only in the cases $k = -2, -\frac{1}{2}, 1$.

If $k = -2$, then $\beta = -2\alpha$, whence $\gamma = -\alpha - \beta = -\alpha + 2\alpha = \alpha$.

If $k = -\frac{1}{2}$, then $\beta = -\frac{1}{2}\alpha$, whence $\gamma = -\alpha + \frac{1}{2}\alpha = -\frac{1}{2}\alpha = \beta$.

If $k = 1$, then $\beta = \alpha$.

This completes the proof.   □

4 The Theorem on the Discriminant Revisited

After presenting the proof of the previous section in class, I switched history textbooks and discovered that Victor Katz gives as an exercise the proof that, in the case where p, q are negative, if the discriminant is negative then there are three real roots. This suggested that there had to be a simpler, more direct proof of the full Theorem. I came up with the following using a bit of Calculus.

Let
$$P(X) = X^3 + pX + q, \quad D = q^2 + \frac{4}{27}p^3.$$

Differentiation yields
$$P'(X) = 3X^2 + p.$$

If $p > 0$, then $P'(x) > 0$ for all values of x, whence P has an unique real root. Note too that in this case $D > 0$.

If $p = 0$, then $P(X) = X^3 + q$ and either $D = q^2 = 0$ and P has a three-fold repeated root at 0 or $D = q^2 > 0$ and P has one real and two complex roots, namely the three cube roots of $-q$.

The heart of the proof is the case in which $p < 0$. In this case, P' has two distinct real roots,
$$x = \pm\sqrt{\frac{-p}{3}}.$$

Now $P''(X) = 6X$, whence

$$P''\left(\sqrt{\frac{-p}{3}}\right) > 0 \text{ and } P \text{ has a local minimum at } \sqrt{\frac{-p}{3}}$$

$$P''\left(-\sqrt{\frac{-p}{3}}\right) < 0 \text{ and } P \text{ has a local maximum at } -\sqrt{\frac{-p}{3}}$$

The values of these local extrema are

$$\left.\begin{array}{l}
P\left(\sqrt{\frac{-p}{3}}\right) = \frac{-p}{3}\sqrt{\frac{-p}{3}} + p\sqrt{\frac{-p}{3}} + q = \frac{2p}{3}\sqrt{\frac{-p}{3}} + q \\[2mm]
P\left(-\sqrt{\frac{-p}{3}}\right) = \frac{p}{3}\sqrt{\frac{-p}{3}} - p\sqrt{\frac{-p}{3}} + q = -\frac{2p}{3}\sqrt{\frac{-p}{3}} + q.
\end{array}\right\} \quad (16)$$

We now consider what happens when $D = 0, D < 0$, and $D > 0$.

We begin with the simplest case: $D = 0$. Then

$$q^2 = -\frac{4}{27}p^3 \tag{17}$$

$$|q| = 2\left(\frac{-p}{3}\right)\sqrt{\frac{-p}{3}} = \frac{-2p}{3}\sqrt{\frac{-p}{3}}.$$

Note that $q \neq 0$, for otherwise, by (17), we have $p = 0$, contrary to our assumption $p < 0$.

If $q > 0$, we have

$$q = |q| = \frac{-2p}{3}\sqrt{\frac{-p}{3}},$$

whence

$$P\left(\sqrt{\frac{-p}{3}}\right) = \frac{2p}{3}\sqrt{\frac{-p}{3}} + q = -q + q = 0,$$

and P has a repeated root at $\sqrt{\frac{-p}{3}}$. Similarly, if $q < 0$,

$$-q = |q| = \frac{-2p}{3}\sqrt{\frac{-p}{3}},$$

whence

$$P\left(-\sqrt{\frac{-p}{3}}\right) = -\frac{2p}{3}\sqrt{\frac{-p}{3}} + q = -q + q = 0,$$

and P has a repeated root at $-\sqrt{\frac{-p}{3}}$.

In the subcase in which $D < 0$, we have

$$q^2 < \frac{-4}{27}p^3,$$

whence

$$|q| < \frac{-2p}{3}\sqrt{\frac{-p}{3}}.$$

Thus

$$P\left(\sqrt{\frac{-p}{3}}\right) = \frac{2p}{3}\sqrt{\frac{-p}{3}} + q \leq \frac{2p}{3}\sqrt{\frac{-p}{3}} + |q| < 0 <$$

$$< \frac{-2p}{3}\sqrt{\frac{-p}{3}} - |q| \leq \frac{-2p}{3}\sqrt{\frac{-p}{3}} + q = P\left(-\sqrt{\frac{-p}{3}}\right),$$

and P has a root between $-\sqrt{\frac{-p}{3}}$ and $\sqrt{\frac{-p}{3}}$. Moreover, for large positive values of x, $P(x)$ is positive, whence P has a root to the right of $\sqrt{\frac{-p}{3}}$. Similarly, P has a root to the left of $-\sqrt{\frac{-p}{3}}$. In all, P has three distinct real roots.

Finally, consider the subcase where $D > 0$. Here we have

$$q^2 > \frac{-4}{27}p^3 > 0,$$

$$|q| > \frac{-2p}{3}\sqrt{\frac{-p}{3}}.$$

If $q > 0$,

$$q = |q| > \frac{-2p}{3}\sqrt{\frac{-p}{3}}$$

and

$$P\left(\sqrt{\frac{-p}{3}}\right) = \frac{2p}{3}\sqrt{\frac{-p}{3}} + q > 0.$$

As the relative minimum is positive, P has only one real root.

If $q < 0$,

$$-q = |q| > \frac{-2p}{3}\sqrt{\frac{-p}{3}}$$

and

$$P\left(-\sqrt{\frac{-p}{3}}\right) = \frac{-2p}{3}\sqrt{\frac{-p}{3}} + q < 0.$$

In this case the relative maximum is negative and P has only one real root.

Again, q cannot be 0 in the case where $p < 0$, and we have completed the proof.

Addendum

A variant of the above proof can be given by which one argues, in the case $p < 0$, from the number of real roots P has to the sign of the discriminant instead of *vice versa*.

Suppose P has three distinct roots, $\alpha < \beta < \gamma$. By Rolle's Theorem, we have

$$\alpha < -\sqrt{\frac{-p}{3}} < \beta < \sqrt{\frac{-p}{3}} < \gamma.$$

By the Intermediate Value Theorem, we also know P to be

negative on $(-\infty, \alpha)$

positive on (α, β)

negative on (β, γ)

positive on (γ, ∞).

In particular, using (16), we see

$$P\left(-\sqrt{\frac{-p}{3}}\right) = -\frac{2p}{3}\sqrt{\frac{-p}{3}} + q > 0, \tag{18}$$

$$P\left(\sqrt{\frac{-p}{3}}\right) = \frac{2p}{3}\sqrt{\frac{-p}{3}} + q < 0. \tag{19}$$

If $q \leq 0$, (18) yields

$$-\frac{2p}{3}\sqrt{\frac{-p}{3}} > -q \geq 0,$$

whence squaring yields

$$\frac{4p^2}{9}\left(\frac{-p}{3}\right) > q^2,$$

i.e.

$$0 > q^2 + \frac{4}{27}p^3 = D.$$

If $q > 0$, (19) yields

$$0 < q < -\frac{2p}{3}\sqrt{\frac{-p}{3}},$$

whence

$$q^2 < -\frac{4}{27}p^3,$$

i.e.

$$D = q^2 + \frac{4}{27}p^3 < 0.$$

Should P have repeated real roots, Rolle's Theorem yields one of the possibilities:

$$\alpha = -\sqrt{\frac{-p}{3}} = \beta < \sqrt{\frac{-p}{3}} < \gamma,$$

or

$$\alpha < -\sqrt{\frac{-p}{3}} < \beta = \sqrt{\frac{-p}{3}} = \gamma,$$

or

$$\alpha = -\sqrt{\frac{-p}{3}} = \beta = \sqrt{\frac{-p}{3}} = \gamma.$$

In any case, we appeal once again to (16) to conclude that for some choice of sign we have

$$\frac{\pm 2p}{3}\sqrt{\frac{-p}{3}} + q = 0,$$

whence isolating q and squaring both sides of the equation yields

$$\frac{-4p^3}{27} = q^2,$$

and $D = \frac{4}{27}p^3 + q^2 = 0$.

If P has an unique real root, say α, Rolles's Theorem tells us P is

$$\text{negative on } (-\infty, \alpha)$$
$$\text{positive on } (\alpha, \infty).$$

Because P is decreasing on the interval from $-\sqrt{-p/3}$ to $\sqrt{-p/3}, \alpha$ cannot lie in that interval, else P goes from positive to negative. Thus, either both extrema occur to the left of α and yield negative values, or both occur to the right and yield positive values. In the first case, consider

$$P\left(-\sqrt{\frac{-p}{3}}\right) = -\frac{2p}{3}\sqrt{\frac{-p}{3}} + q < 0.$$

Here,

$$0 < -\frac{2p}{3}\sqrt{\frac{-p}{3}} < -q$$

and squaring yields

$$-\frac{4}{27}p^3 < q^2,$$

and $D > 0$. In the second case, consider

$$P\left(\sqrt{\frac{-p}{3}}\right) = \frac{2p}{3}\sqrt{\frac{-p}{3}} + q > 0.$$

Now we have

$$q > -\frac{2p}{3}\sqrt{\frac{-p}{3}} > 0$$

and we can square again to conclude $D > 0$.

5 Computational Considerations

Let us begin with Lǐ Yě's equation for Qín Jiǔsháo's problem of the circular city, the tree to the north, and the observer east of the south gate.

5.1 Example. $X^3 + 3X^2 - 972 = 0$.

The unique real solution is $X = 9$. The Chinese would have solved this by an approximation technique, which would quickly have yielded the exact value. Today we would probably use a graphical calculator, which might or might not yield the exact value but which would yield a value so close to 9 that we might go ahead and evaluate the polynomial at 9 just to see if it worked.

Because we like exact solutions, we might first check for rational solutions, which in this case would be integral divisors of 972. As 972 has 18 positive integral divisors and 36 in all, this task might seem a bit daunting until we realise that today's pocket calculators can handle the task without programming quite easily. On the *TI-83+* one first goes into the equation editor and enters the polynomial

$$Y_1 = X^\wedge 3 + 3X^\wedge 2 - 972.$$

One then exits the equation editor and creates a list of positive divisors of 972

$$\{1, 2, 3, 4, 6, 9, 12, 18, 27, 36, 54, 81, 108, 162, 243, 324, 486, 972\} \to L_1$$

One then makes a list of negative divisors,

$$-L_1 \to L_2,$$

and combines the two lists,

$$\text{augment}\,(L_1, L_2) \to L_1.$$

One then evaluates the polynomial at all points on the list,

$$Y_1\,(L_1) \to L_2.$$

One can now open the List Editor and scroll down until a 0 is found in L_2 and read off the accompanying value of X in L_1.

The grubby part of the preceding is creating the list L_1. However, we do not even have to do this by hand. 972 is quickly factored: $972 = 2^2 3^5$. One can start with the list of powers of 3 dividing 972:

$$\{1, 3, 9, 27, 81, 243\} \to L_1,$$

and then construct the lists:

$$2L_1 \to L_2$$
$$2L_2 \to L_3$$
$$\text{augment}\,(L_2, L_3) \to L_2$$
$$\text{augment}\,(L_1, L_2) \to L_1.$$

The list L_1 now contains, in some order, all the positive divisors of 972. One can now evaluate P at all points in L_1 and $-L_1$ as before. One might first want to apply the command

$$\text{ClrList}\ L_3$$

before opening the List Editor to avoid any distraction in searching for the 0 in L_2 in the Editor.

Having done all of this, one finds 9 as the sole rational solution to Li Yeh's equation. Dividing the polynomial by $X - 9$ produces a quadratic factor,

$$X^2 + 12X + 108,$$

the roots of which are

$$X = \frac{-12 \pm \sqrt{144 - 432}}{2} = \frac{-12 \pm \sqrt{-228}}{2}$$
$$= \frac{-12 \pm 12\sqrt{-2}}{2} = -6 \pm 6\sqrt{-2},$$

both complex.

The Renaissance algebraists, however, would not have had pocket calculators operating on lists of arguments and had to do all of their calculations by hand. I haven't checked, but the realisation that the rational solutions to Lĭ Yĕ's equation are divisors of 972, being number-theoretic in character, strikes me as belonging to a later period. Thus, they would need Tartaglia's formula.

The substitution $X = Y - 1$ results in the equation,

$$Y^3 - 3Y - 970 = 0. \tag{20}$$

Since $Y = X + 1$, the root $X = 9$ is transformed into $Y = 10$. Here

$$p = -3, \quad q = -970, \quad D = 970^2 + \frac{4}{27}(-3)^3 = 970^2 - 4$$

$$a^3 = \frac{970 \pm \sqrt{970^2 - 4}}{2} = 485 \pm 198\sqrt{6}$$
$$= 969.9989691 \text{ or } .0010309289$$

on the calculator. Thus

$$a = \sqrt[3]{485 \pm 198\sqrt{6}}$$
$$= 9.898979486 \text{ or } .1010205144$$

on the calculator. Now $b = \frac{p}{3a} = \frac{-3}{3a} = -\frac{1}{a}$, whence

$$Y = a - b = a + \frac{1}{a} = 10$$

for either value of a. But is this approximate or exact? We can verify that it is exact by plugging 10 into (20). If, however, we round a^3 to only 4 decimals,

$$a^3 = 969.9990,$$

then we get

$$a + \frac{1}{a} = 10.0000001.$$

Of course, if we round again to 4 decimals, we get 10, but we might be less inclined to consider it exact.

We can verify that the Tartaglia-Cardano solution

$$Y = \sqrt[3]{485 + 198\sqrt{6}} - \sqrt[3]{-485 + 198\sqrt{6}}$$

is 10 without resorting to approximation via the calculator by doing some extra algebra.

Starting with 10 and observing that

$$a + \frac{1}{a} = 10,$$

we obtain a quadratic equation,

$$a^2 - 10a + 1 = 0,$$

with solutions

$$a = \frac{10 \pm \sqrt{100 - 4}}{2} = 5 \pm 2\sqrt{6}.$$

Cubing these yields

$$\left(5 + 2\sqrt{6}\right)^3 = 485 + 198\sqrt{6}$$

$$\left(5 - 2\sqrt{6}\right)^3 = 485 - 198\sqrt{6}.$$

So, choosing, e.g., $a^3 = 485 + 198\sqrt{6}$, we have

$$a = 5 + 2\sqrt{6}, \quad b = \frac{-1}{5 + 2\sqrt{6}} = -5 + 2\sqrt{6}$$

and

$$Y = 5 + 2\sqrt{6} - \left(-5 + 2\sqrt{6}\right) = 10.$$

There is another method that applies here, one that doesn't require us to start with 10, but which requires a bit of theory for justification and which does not always work. The solution to the equation,

$$Z^3 - \left(485 + 198\sqrt{6}\right) = 0,$$

is an *algebraic integer*. Now *if* a solution already exists in $\mathbb{Q}\left(\sqrt{6}\right)$, then the solution is of the form $m + n\sqrt{6}$, with m, n integral. So one solves

$$\left(m + n\sqrt{6}\right)^3 = m^3 + 3m^2 n\sqrt{6} + 3m6n^2 + n^3 6\sqrt{6}.$$

In particular,

$$m^3 + 18mn^2 = 485,$$

i.e.,

$$m\left(m^2 + 18n^2\right) = 485,$$

and m must be a divisor of 485. Again, to test the divisors, we can resort to the *TI-83+*:

$$\{1, 5, 97, 485\} \to L_1$$
$$-L_1 \to L_2$$
$$\text{augment}\,(L_1, L_2) \to L_1.$$

L_1 is now the list of divisors of 485. Now overwrite L_2:

$$\left(485 \div L_1 - L_1^2\right) \div 18 \to L_2.$$

One can now enter the List Editor to look for perfect integral squares n^2 in L_2 and their corresponding m-values in L_1. Or, if one is not good at recognising perfect squares, one can take square roots:

$$\sqrt{(L_2)} \to L_2,$$

making sure the calculator is in complex mode to avoid error messages when taking square roots of the negative entries in L_2. The only integer in the square root list is 2, corresponding to $m = 5$. This yields $m = 5, n = \pm 2$ as candidates and, as above, both work and yield $Y = 10$.

At some point one must again mention that 10 is the only real root. We can see this by dividing the polynomial by $Y - 10$ and solving the resulting quadratic or by noting that $D > 0$.

5.2 Example. $X^3 + 2X^2 - X - 2 = 0.$

This equation is easily solved by conventional methods, perhaps most easily by factoring-by-grouping:

$$\begin{aligned}
X^3 + 2X^2 - X - 2 &= \left(X^3 + 2X^2\right) + (-X - 2) \\
&= X^2\,(X + 2) - (X + 2) \\
&= \left(X^2 - 1\right)(X + 2) \\
&= (X + 1)\,(X - 1)\,(X + 2),
\end{aligned}$$

whence $X = \pm 1, -2$ are the solutions. Let us see how the general solution behaves in this example.

The substitution $X = Y - \frac{2}{3}$ produces

$$Y^3 - \frac{7}{3}Y - \frac{20}{27} = 0,$$

with solutions $Y = -1 + \frac{2}{3}, 1 + \frac{2}{3}, -2 + \frac{2}{3} = \frac{-1}{3}, \frac{5}{3}, \frac{-4}{3}$. We have

$$p = \frac{-7}{3}, \quad q = \frac{-20}{27}, \quad D = \frac{400}{729} + \frac{4}{27}\left(\frac{-343}{27}\right) = \frac{-972}{729}$$

$$a^3 = \frac{\frac{20}{27} \pm \sqrt{\frac{-972}{729}}}{2} = \frac{10 \pm 9\sqrt{-3}}{27}.$$

The denominator of a^3 being a perfect cube, we need only find a cube root of the numerator to find $3a$ and thus a.

Again we consider

$$Z^3 - \left(10 + 9\sqrt{-3}\right) = 0.$$

If a solution to this exists in $\mathbb{Q}\left(\sqrt{-3}\right)$, then there is an algebraic integer in $\mathbb{Q}\left(\sqrt{-3}\right)$ satisfying the equation. The algebraic integers of $\mathbb{Q}\left(\sqrt{-3}\right)$, however, are no longer only of the form $m + n\sqrt{-3}$ for integral m, n, but also include numbers of the form $\frac{m+n\sqrt{-3}}{2}$. [Basically, the reason is this: The algebraic integers in fields $\mathbb{Q}\left(\sqrt{D}\right)$ are those numbers $r + s\sqrt{D}$, with r, s rational, which satisfy equations

$$X^2 + BX + C = 0, \quad B, C \text{ integers.}$$

These solutions are of the form $\frac{-B\pm\sqrt{B^2-4C}}{2}$. To be in $\mathbb{Q}\left(\sqrt{D}\right)$, $B^2 - 4C$ must be a square or a square multiplied by D. If B is even, the denominator will cancel out and the solutions will be of the form $m + n\sqrt{D}$. If B is odd, after factoring out the odd square from $B^2 - 4C$, D is represented as $\left(B'\right)^2 - 4C'$. If D is congruent to 3 mod 4 or if D is even, this will not happen; but if D is congruent to 1 mod 4, as is the case for $D = -3$, this does occur and one is stuck with the 2 in the denominator.]

Thus, we are looking for integers m, n for which

$$\left(\frac{m + n\sqrt{-3}}{2}\right)^3 = 10 + 9\sqrt{-3},$$

i.e.

$$\frac{1}{8}\left(m^3 + 3\sqrt{-3}m^2n - 9mn^2 - 3\sqrt{-3}n^3\right) = 10 + 9\sqrt{-3}.$$

In particular, we want

$$\frac{1}{8}\left(m^3 - 9mn^2\right) = 10,$$

i.e.,

$$m\left(m^2 - 9n^2\right) = 80.$$

Thus, we must test all divisors m of 80. Playing with the *TI-83+* as in Example 5.2, we obtain the following candidates:

$$m = -5, n = \pm 1$$
$$m = -1, n = \pm 3$$
$$m = -4, n = \pm 2.$$

And, in fact,

$$\left(\frac{5 + \sqrt{-3}}{2}\right)^3 = \left(\frac{-1 - 3\sqrt{-3}}{2}\right)^3 = \left(\frac{-4 + 2\sqrt{-3}}{2}\right)^3 = 10 + 9\sqrt{-3}$$

$$\left(\frac{5 - \sqrt{-3}}{2}\right)^3 = \left(\frac{-1 + 3\sqrt{-3}}{2}\right)^3 = \left(\frac{-4 - 2\sqrt{-3}}{2}\right)^3 = 10 - 9\sqrt{-3}.$$

We may take $3a$ to be any one of these cube roots and divide by 3 to obtain

$$a = \frac{5 \pm \sqrt{-3}}{6} \quad \text{or} \quad \frac{-1 \pm 3\sqrt{-3}}{6} \quad \text{or} \quad \frac{-4 \pm 2\sqrt{-3}}{6},$$

and

$$Y = 2\,Re\,(a) = \frac{10}{6} = \frac{5}{3} \quad \text{or} \quad \frac{-2}{6} = \frac{-1}{3} \quad \text{or} \quad \frac{-8}{6} = \frac{-4}{3}.$$

[If we take one of the values of $3a$ as X, say $X = \frac{5+\sqrt{-3}}{2}$, and square it we obtain

$$X^2 = \frac{11 + 5\sqrt{-3}}{2}.$$

We can remove the square root by subtracting $5X$:

$$X^2 - 5X = 7,$$

whence

$$X^2 - 5X - 7 = 0,$$

and X is indeed an algebraic integer.]

5.3 Example. $Y^3 + Y - 1 = 0$.

This is Example 1 of Section 2. Here we have

$$a^3 = \frac{9 \pm \sqrt{93}}{18}.$$

The denominator is not a perfect cube, but if we multiply both numerator and denominator by 12 it becomes one:

$$a^3 = \frac{108 \pm 12\sqrt{-3}}{2^3 3^3}.$$

As in the last example, 93 is congruent to 1 mod 4 and the algebraic integers of $\mathbb{Q}\left(\sqrt{93}\right)$ are of the form $\frac{m+n\sqrt{93}}{2}$, for integers m, n. Thus we wish to solve

$$\left(\frac{m + n\sqrt{93}}{2}\right)^3 = 108 + 12\sqrt{93}.$$

This leads to

$$\frac{1}{8}\left(m^2 + 3mn^2 93\right) = 108,$$

i.e.

$$m\left(m^2 + 279n^2\right) = 864.$$

Now $864 = 2^5 3^3$ has $(5+1)(3+1) = 24$ positive divisors, 48 divisors in all. Dragging out the *TI-83+* again, we can punch in

$$\{1, 2, 4, 8, 16, 32\} \rightarrow L_1$$
$$3L_1 \rightarrow L_2$$
$$\text{augment}\,(L_1, L_2) \rightarrow L_1$$
$$9L_1 \rightarrow L_2$$
$$\text{augment}\,(L_1, L_2) \rightarrow L_1$$
$$-L_1 \rightarrow L_2$$
$$\text{augment}\,(L_1, L_2) \rightarrow L_1.$$

L_1 now contains, in some order, all 48 divisors of 864. Now calculate

$$\sqrt{\left(\left(864 \div L_1 - L_1^2\right) \div 279\right)} \rightarrow L_2.$$

The only candidates that appear on the list are $m = 3, n = \pm 1$. But

$$\left(\frac{3 + \sqrt{93}}{2}\right)^3 = 108 + 15\sqrt{93}$$

$$\left(\frac{3 - \sqrt{93}}{2}\right)^3 = 108 - 15\sqrt{93},$$

and we conclude that we cannot represent a in the form $\frac{m+n\sqrt{93}}{2}$, i.e. a is not in $\mathbb{Q}\left(\sqrt{93}\right)$.

What went "wrong" in this example is that the equation in question has no rational solution. Recall from the discussion of the angle trisection problem the crucial lemma on cubic polynomials that a polynomial $X^3 + BX^2 + CX + D$ with a constructible solution already has a rational solution. For geometric reasons our definition of constructibility was restricted to real numbers. Algebraically, however, there is no need for such a restriction. If one defines the constructible complex numbers to be those that can be reached by a chain of quadratic extensions of $\mathbb{Q}$ with no requirement on the reality of the root, then one can readily verify that the same proof shows that a cubic $X^3 + BX^2 + CX + D$ with rational coefficients that has no rational root has no constructible complex root. If, in our example, $\sqrt[3]{108 + 12\sqrt{93}}$ were expressible as $\frac{m+n\sqrt{93}}{2}$, or were equal to any other constructible real, then $Y^3 + Y - 1$ would have a rational root.

5.4 Example. $Y^3 - 3Y - 1 = 0$.

This equation comes from Example 2.2 above. It resulted in

$$a^3 = \frac{1 + \sqrt{-3}}{2}.$$

As in Example 6, the present equation has no rational solution and a is not constructible. One can, however, apply DeMoivre's Theorem to calculate a:

$$\begin{aligned} a^3 &= \cos 60 + i \sin 60 \\ &= \cos 420 + i \sin 420 \\ &= \cos 780 + i \sin 780, \end{aligned}$$

which has the three distinct cube roots:

$$\begin{aligned} a_1 &= \cos 20 + i \sin 20 \\ a_2 &= \cos 140 + i \sin 140 \\ a_3 &= \cos 260 + i \sin 260, \end{aligned}$$

resulting in the real roots,

$$\begin{aligned} y_1 &= 2Re\,(a_1) = 2 \cos 20 = 1.879385242 \\ y_2 &= 2Re\,(a_2) = 2 \cos 140 = -1.532088886 \\ y_3 &= 2Re\,(a_3) = 2 \cos 260 = -.3472963553. \end{aligned}$$

Now, DeMoivre wasn't born until 1667 and his formula was not available to the Renaissance mathematicians. But, by the end of the 16th century, Viète was able to apply trigonometry to these problems. Substituting $Y = 2Z$ transforms the equation into

$$8Z^3 - 6Z - 1 = 0,$$

i.e.

$$4Z^3 - 3Z = \frac{1}{2}.$$

The coefficients ought to be familiar enough to suggest setting $Z = \cos \theta$:

$$\cos 3\theta = 4 \cos^3 \theta - 3 \cos \theta = \frac{1}{2},$$

resulting in $3\theta = 60\,, 420,\ 780$ and thus $Y = 2Z = 2 \cos 20, 2 \cos 120, 2 \cos 260$, as before.

Viète's trigonometric approach always works when the discriminant is negative:

5.5 Theorem. *Let p, q be real numbers with*

$$D = q^2 + \frac{4}{27}p^3 < 0. \tag{21}$$

Then

$$\frac{2\sqrt{-p}}{\sqrt{3}} \cos\left[\frac{1}{3} \cos^{-1}\left(\frac{\sqrt{27}q}{2p\sqrt{-p^3}}\right)\right]$$

is a real solution to

$$X^3 + pX + q = 0.$$

Proof. Substituting $X = \frac{2\sqrt{-p}}{\sqrt{3}}Y$ into the equation results in

$$\frac{-8p\sqrt{-p}}{3\sqrt{3}}Y^3 + \frac{2p\sqrt{-p}}{\sqrt{3}}Y = -q,$$

and thus

$$4Y^3 - 3Y = \frac{3\sqrt{3}q}{2p\sqrt{-p}}. \tag{22}$$

Now the square of the right-hand side of this is

$$-\frac{27q^2}{4p^3},$$

which is less than 1 in absolute value by (21):

$$q^2 < -\frac{4}{27}p^3$$

$$\frac{27q^2}{-4p^3} < 1,$$

since $-p^3 > 0$. Thus $\frac{q\sqrt{27}}{2p\sqrt{-p}}$ is the cosine of some angle θ, whence (22) is indeed of the form

$$4Y^3 - 3Y = \cos\theta,$$

and $Y = \cos(\theta/3)$ is a solution. The Theorem follows.   □

Addendum

Note that Viète's trick accounts for the three distinct roots when $D < 0$. His substitution and a companion actually account for all these cases. Let the equation $X^3 + pX + q = 0$ be given and assume, to avoid the trivial cases, that $p \neq 0$.

If $p < 0$, Viète's substitution $X = \frac{2\sqrt{-p}}{\sqrt{3}}Y$ results in an equation,

$$4Y^3 - 3Y = \alpha, \quad \text{where } \alpha = \frac{3\sqrt{3}q}{2p\sqrt{-p}}. \tag{23}$$

Now we easily see that

$$|\alpha| < 1 \quad \text{when} \quad D < 0$$
$$|\alpha| = 1 \quad \text{when} \quad D = 0$$
$$|\alpha| > 1 \quad \text{when} \quad D > 0.$$

A glance at the curve $W = 4Y^3 - 3Y$ is enough to convince one that $4Y^3 - 3Y = \alpha$ has

3 distinct real solutions when $|\alpha| < 1$, i.e. when $D < 0$
3 real solutions, 1 repeated when $|\alpha| = 1$, i.e. when $D = 0$
an unique real solution when $|\alpha| > 1$, i.e. when $D > 0$.

To obtain a rigorous proof, set $f(y) = 4y^3 - 3y$ and look at $f'(y) = 12y^2 - 3$ to conclude f increases on $(-\infty, -1/2]$ from $-\infty$ to 1, decreases on $[-1/2, 1/2]$ from 1 to -1, and increases on $[1/2, +\infty)$ from -1 to $+\infty$.

If $p > 0$, the related substitution $X = \frac{2\sqrt{p}}{\sqrt{3}} Y$ results in an equation,

$$4Y^3 + 3Y = \alpha, \quad \text{where } \alpha = \frac{3\sqrt{3}q}{2p\sqrt{p}}.$$

D is greater than 0 and $f(y) = 4y^3 + 3y$ is one-to-one because $f'(y) = 12y^2 + 3 > 0$ for all values of y.

The two substitutions are not obvious ones to make and one does use the Calculus in this proof, but it is by far the simplest of those presented here. It also allows one to find the solution to the cubic without recourse to complex numbers, but at the cost of introducing transcendental functions. We have already seen this with the cosine in the case when $D < 0$. The treatment carries over unchanged when $D = 0$, i.e. $|\alpha| = 1$.

When $D > 0$, one can replace the use of the cosine by application of the hyperbolic functions,

$$\cosh x = \frac{e^x + e^{-x}}{2}, \quad \sinh x = \frac{e^x - e^{-x}}{2}.$$

Here the "triple angle" formulæ are

$$\cosh 3x = 4 \cosh^3 x - 3 \cosh x$$
$$\sinh 3x = 4 \sinh^3 x + 3 \sinh x.$$

If $p < 0$, equation (23) reads

$$4Y^3 - 3Y = \alpha, \quad \text{with } |\alpha| > 1.$$

If $\alpha > 1$, the unique solution is given by

$$Y = \cosh\left(\frac{1}{3} \cosh^{-1} \alpha\right),$$

while if $\alpha < -1$, it is

$$Y = -\cosh\left(\frac{1}{3}\cosh^{-1}(-\alpha)\right).$$

If $p > 0$, every α has an inverse hyperbolic sine, and we can take

$$Y = \sinh\left(\frac{1}{3}\sinh^{-1}\alpha\right).$$

6 One Last Proof

There is another proof of the theorem on the discriminant, one that uses no Calculus, no trigonometry, and no hyperbolic functions. However, i) it uses some sneaky calculations, and ii) its motivation uses more and deeper algebra.

One begins with an arbitrary (not necessarily reduced) cubic polynomial with (possibly complex) roots r_1, r_2, r_3 and considers the Vandermonde determinant

$$\Delta = \begin{vmatrix} 1 & r_1 & r_1^2 \\ 1 & r_2 & r_2^2 \\ 1 & r_3 & r_3^2 \end{vmatrix} = \prod_{i<j}(r_j - r_i), \tag{24}$$

and squares it:

$$\Delta^2 = \prod_{i<j}(r_j - r_i)^2. \tag{25}$$

The quantity Δ^2 is sometimes taken to be the discriminant of the equation because, like D, it discriminates the various cases:

6.1 Lemma. *Let P be a cubic polynomial with real coefficients and roots $r_1, r_2, r_3 \in \mathbb{C}$, and let Δ^2 be as in (25). Then*
i. P has only one real root iff $\Delta^2 < 0$
ii. P has 3 real but not distinct roots iff $\Delta^2 = 0$
iii. P has 3 distinct real roots iff $\Delta^2 > 0$.

Proof. Since the three conditions on Δ^2 are exhaustive, it suffices to prove the left-to-right implications.

i. If P has one real root r and two complex roots $a \pm bi$ for $a, b \in \mathbb{R}$, then by (24)

$$\Delta = (r - a - bi)(r - a + bi)(a + bi - a + bi) = \left((r-a)^2 + b^2\right) \cdot 2bi,$$

whence

$$\Delta^2 = -4b^2\left((r-a)^2 + b^2\right)^2 < 0.$$

ii. If two roots r_i and r_j are identical, then one of the factors defining Δ is 0, whence $\Delta^2 = 0$.

iii. If r_1, r_2, r_3 are distinct real numbers, then by (25), Δ^2 is the product of 3 nonzero squares, whence is positive.   □

The proof of this lemma is infinitely simpler than the proofs given for D. However, D can be calculated directly from the coefficients of $P(X)$ when P is reduced. For Δ^2, it appears we have to know the roots first. This is only an illusion, for Δ^2 can be found directly from the coefficients of P and, when P is reduced, i.e. of the form $X^3 + pX + q$, Δ^2 is just $-27D$. Verifying this is where things get grubby.

First, let us do the easy part. Going back to the square of the Vandermonde,

$$
\Delta^2 = \begin{vmatrix} 1 & r_1 & r_1^2 \\ 1 & r_2 & r_2^2 \\ 1 & r_3 & r_3^2 \end{vmatrix} \cdot \begin{vmatrix} 1 & r_1 & r_1^2 \\ 1 & r_2 & r_2^2 \\ 1 & r_3 & r_3^2 \end{vmatrix}
$$

$$
= \begin{vmatrix} \begin{bmatrix} 1 & 1 & 1 \\ r_1 & r_2 & r_3 \\ r_1^2 & r_2^2 & r_3^2 \end{bmatrix} \cdot \begin{bmatrix} 1 & r_1 & r_1^2 \\ 1 & r_2 & r_2^2 \\ 1 & r_3 & r_3^2 \end{bmatrix} \end{vmatrix}
$$

$$
= \begin{vmatrix} 3 & \sum r_i & \sum r_i^2 \\ \sum r_i & \sum r_i^2 & \sum r_i^3 \\ \sum r_i^2 & \sum r_i^3 & \sum r_i^4 \end{vmatrix} \tag{26}
$$

The entries in the determinant are *symmetric polynomials* in r_1, r_2, r_3. That is, each of these sums is a polynomial σ for which

$$
\sigma(r_1, r_2, r_3) = \sigma(r_{\tau 1}, r_{\tau 2}, r_{\tau 3})
$$

for every permutation τ of $\{1, 2, 3\}$. A theorem going back to Newton tells us that every symmetric polynomial in r_1, r_2, r_3 can be expressed in terms of the *elementary symmetric polynomials*,

$$
\sigma_0 = 1
$$
$$
\sigma_1 = r_1 + r_2 + r_3
$$
$$
\sigma_2 = r_1 r_2 + r_2 r_3 + r_3 r_1
$$
$$
\sigma_3 = r_1 r_2 r_3.
$$

Moreover, but for the signs, these are the coefficients of P:

$$
P(X) = (X - r_1)(X - r_2)(X - r_3) = \sigma_0 X^3 - \sigma_1 X^2 + \sigma_2 X - \sigma_3.
$$

Thus, Δ^2 can be expressed in terms of the coefficients of P. We shall verify this for the case in which $P(X) = X^3 + pX + q$ is reduced.

The first thing to notice is that $\sigma_0, \sigma_1, \sigma_2$, and σ_3 are simply expressed in terms of p and q:

$$
\sigma_0 = 1, \quad \sigma_1 = 0, \quad \sigma_2 = p, \quad \sigma_3 = -q.
$$

The entries in the determinant to be further expressed in terms of p, q are $\sum r_i^2, \sum r_i^3$, and $\sum r_i^4$. For the first of these, notice that

$$(r_1 + r_2 + r_3)^2 = r_1^2 + r_2^2 + r_3^2 + 2r_1r_2 + 2r_2r_3 + 2r_3r_1,$$

i.e., $0 = \sum r_i^2 + 2p$, whence

$$\sum r_i^2 = -2p. \tag{27}$$

To find $\sum r_i^3$, note that

$$\left(r_1^2 + r_2^2 + r_3^2\right)(r_1 + r_2 + r_3) = r_1^3 + r_2^3 + r_3^3 +$$
$$r_1^2 r_2 + r_1^2 r_3 + r_2^2 r_1 + r_2^2 r_3 + r_3^2 r_1 + r_3^2 r_2$$
$$= r_1^3 + r_2^3 + r_3^3 +$$
$$r_1 r_2(r_1 + r_2) + r_1 r_3(r_1 + r_3) + r_2 r_3(r_2 + r_3)$$
$$= r_1^3 + r_2^3 + r_3^3 +$$
$$r_1 r_2(r_1 + r_2 + r_3) - r_1 r_2 r_3 +$$
$$r_1 r_3(r_1 + r_2 + r_3) - r_1 r_2 r_3 +$$
$$r_2 r_3(r_1 + r_2 + r_3) - r_1 r_2 r_3,$$

whence $0 = \sum r_i^3 + 3q$, i.e.

$$\sum r_i^3 = -3q. \tag{28}$$

Finally, to find $\sum r_i^4$, consider

$$\left(r_1^3 + r_2^3 + r_3^3\right)(r_1 + r_2 + r_3) = r_1^4 + r_2^4 + r_3^4 +$$
$$r_1^3 r_2 + r_1^3 r_3 + r_2^3 r_1 + r_2^3 r_3 + r_3^3 r_1 + r_3^3 r_2$$
$$= r_1^4 + r_2^4 + r_3^4 +$$
$$r_1 r_2(r_1^2 + r_2^2) + r_1 r_3(r_1^2 + r_3^2) + r_2 r_3(r_2^2 + r_3^2)$$
$$= r_1^4 + r_2^4 + r_3^4 +$$
$$r_1 r_2(r_1^2 + r_2^2 + r_3^2) - r_1 r_2 r_3^2 +$$
$$r_1 r_3(r_1^2 + r_2^2 + r_3^2) - r_1 r_2^2 r_3 +$$
$$r_2 r_3(r_1^2 + r_2^2 + r_3^2) - r_1^2 r_2 r_3$$
$$= \sum r_i^4 + \left(\sum r_i r_j\right)\left(\sum r_i^2\right) - r_1 r_2 r_3\left(\sum r_i\right),$$

whence $0 = \sum r_i^4 + p(-2p) + q \cdot 0$, i.e.

$$\sum r_i^4 = 2p^2. \tag{29}$$

Using the values (27), (28), and (29) in the expression (26) for Δ^2 gives

$$\Delta^2 = \begin{vmatrix} 3 & 0 & -2p \\ 0 & -2p & -3q \\ -2p & -3q & 2p^2 \end{vmatrix} = 3\begin{vmatrix} -2p & -3q \\ -3q & 2p^2 \end{vmatrix} - 2p\begin{vmatrix} 0 & -2p \\ -2p & -3q \end{vmatrix}$$
$$= 3(-4p^3 - 9q^2) - 2p(-4p^2) = -12p^3 - 27q^2 + 8p^3$$
$$= -4p^3 - 27q^2 = -27(q^2 + \frac{4}{27}p^3) = -27D,$$

and Lemma 6.1 yields once again the theorem on the discriminant.

As I said earlier, any History of Mathematics course in which the solution to the cubic equation is given should include a proof that when all three roots are real and distinct the discriminant is negative. I have offered 4 proofs to choose from, the choice depending on various factors— whether the proof given should depend on techniques of the Calculus, whether or not it should appear natural (especially as with the proof of section 4) or may be allowed the introduction of such tricks as the special substitutions in the proof in the addendum to section 5, and whether or not one is planning to discuss a lot of algebra (making the proof of the present section particularly relevant). I, myself, like the proof of section 3 because it has a certain flow to it: after observing that D is a function of only two of the roots, it is natural to look at that function; in this day and age that means graphing it in a 3D program; the radial symmetry of the graph suggests the proof of Lemma 3.2; and the rest follows naturally. If one is willing to use the Calculus, the proof of section 4 is, however, preferable.

Horner's Method

1 Horner's Method

One book that every student of the History of Mathematics ought to be made aware of, even though it is not strictly speaking central to the subject either as a work of mathematics or as one of the history thereof, is Augustus de Morgan's *A Budget of Paradoxes*. The first edition was edited by de Morgan's widow Sophia in 1872, and the second in 1915 by David Eugene Smith. It is a book worth dipping into now and then. In connexion with the work of Qín Jiǔsháo, I looked up what de Morgan had to say about Horner's method.

In his edition, Smith added a footnote giving some background information:

> Davies Gilbert presented the method to the Royal Society in 1819, and it was reprinted in the *Ladies' Diary* in 1838, and in the *Mathematician* in 1843. The method was original as far as Horner was concerned, but it was practically identical with the one used by the Chinese algebraist Ch'in Chiu-shang, in his *Su-shu Chiu-chang* of 1247. But even Ch'in Chiu-shang can hardly be called the discoverer of the method since it is merely the extension of a process for root extracting that appeared in the *Chiu-chang Suan-shu* of the second century B.C.

In his *Source Book in Mathematics,* first published in 1929 and the only one of the major English language source books to include Horner, Smith says of Horner and his method,

> His method closely resembles that which seems to have been developed during the thirteenth century by the Chinese and perfected by Chin Kiu-shao about 1250. It is also very similar to the approximation process effected in 1804 by Paolo Ruffini (1765 - 1822). The probability is, however, that neither Horner nor Ruffini knew of the work of the other and that neither was aware of the ancient Chinese method.

Apparently Horner knew very little of any previous work in approximation as he did not mention in his article the contributions of Vieta, Harriot, Oughtred, or Wallis.

A bit later, he gives the date of publication in *The Mathematician* as 1845[1] and adds that this publication and another in Leybourn's *The Mathematical Repository* in 1830 are revised versions of the paper using ordinary algebra and giving a simpler explanation of the process. In his initial paper, Horner used Taylor's Theorem and hinted that the method applied to transcendental functions as well. In her article on William George Horner in the *Dictionary of Scientific Biography,* Margaret Baron counters this last, saying "his claims that it extended to irrational and transcendental equations were unfounded."

Baron informs us that "the numerical solution of equations was a popular subject in the early nineteenth century" and that

> Horner found influential sponsors in J.R. Young of Belfast and Augustus de Morgan, who gave extracts and accounts of the method in their own publications. In consequence of the wide publicity it received, Horner's method spread rapidly in England but was little used elsewhere in Europe.
>
> Throughout the nineteenth and early twentieth centuries Horner's method occupied a prominent place in standard English and American textbooks on the theory of equations, although, because of its lack of generality, it has found little favor with modern analysts. With the development of computer methods the subject has declined in importance, but some of Horner's techniques have been incorporated in courses in numerical analysis.

Textbooks on the history of mathematics cover a lot of material and may or may not have much to say about Horner, his method, or why de Morgan was so keen on it. My most recent such textbooks are the sixth edition of Burton and the first edition of Katz, both cited in the Bibliography. Burton does not explain the method and states merely that Chu Shih-chieh (fl. 1280 - 1303) applied the technique around 1299— no mention is made of Qín in connexion with this technique. Horner is not in the index to Katz, but Qín's application of it to the equation

$$-X^4 + 763200X^2 - 40642560000 = 0, \tag{1}$$

is worked out thoroughly and the remark that Horner came up with an "essentially similar method in 1819" is made. Equation (1), by the way, is treated whenever one chooses to present the details of Qín's method. Whether it is because this is the example most thoroughly explained by Qín or the first or only one to be translated completely into a Western language I cannot say.

So what is this tool called Horner's Method, due originally to Qín/Ch'in or some predecessors? Why did de Morgan like it, and what did he have to

---

[1] The earlier cited date of 1843 agrees with the *Dictionary of Scientific Biography.*

say about it? Well, popularly conceived it is an algorithm for approximating the solution to a numerical equation. Embedded in it are two subalgorithms— one now called synthetic division can be used to evaluate a polynomial at a given argument, one iterating the application of synthetic division to calculate the coefficients of a new polynomial arising from a given one by performing a simple substitution.

Historical accounts lead one to believe that the difference between Qín's and Horner's presentations of this common method are cosmetic and theoretical. Cosmetically, the differences in approach concern the representation of the equation and the numbers. Qín always arranges his equations so as to make the constant term negative, as in equation (1); Horner, at least in his 1819 paper, would have rewritten (1) as

$$X^4 - 763200X^2 + 40642560000 = 0 \ \text{ or } \ X^4 - 763200X^2 = -40642560000,$$

with the lead coefficient being 1. Qín used counting sticks to represent the coefficients and Horner wrote them out by hand. The significant difference seems to be that Qín explained the working of the method and Horner added a proof that it worked.

I wrote "seems" in connexion with the adjective "significant" in the last line because I am reporting on the impression I received from Smith, Burton, Katz, and similar references, along with such references as Lǐ and Dù's *Chinese Mathematics, A Concise History* and George Gheverghese Joseph's *The Crest of the Peacock,* both of which present Qín's treatment of equation (1). When I look at Horner's 1819 paper, however, more significant differences emerge.

A modern treatment of the method begins with a polynomial,

$$P(X) = a_0 X^n + a_1 X^{n-1} + \ldots + a_{n-1}X + a_n \tag{2}$$

and rewrites it via a partial factoring as follows:

$$P(X) = (\ldots ((a_0 X + a_1)X + a_2)X + \ldots + a_{n-1})X + a_n. \tag{3}$$

The advantage that representation (3) has over (2) in evaluating P at some argument a is immediate: fewer multiplications are involved— and all with the same multiplier a. Indeed, one can evaluate $P(a)$ by generating the short sequence,

$$A_0 = a_0$$
$$A_1 = aA_0 + a_1$$
$$\vdots$$
$$A_{k+1} = aA_k + a_{k+1}$$
$$\vdots$$
$$A_n = aA_{n-1} + a_n,$$

and noting that $A_n = P(a)$.

One usually presents this as a three-row array,

$$
\begin{array}{c|ccccc}
a & a_0 & a_1 \ldots & a_{n-1} & a_n \\
& & aA_0 \ldots & aA_{n-2} & aA_{n-1} \\
\hline
& A_0 & A_1 \ldots & A_{n-1} & A_n
\end{array} .
$$

Here one obtains the entries in the third row by adding the entries immediately above them, while the $(i+1)$-th entry of the second row is obtained by multiplying the i-th entry of the third row by a.

The process just described is called *synthetic division* because the truncated sequence $A_0, A_1, \ldots, A_{n-1}$ is just the sequence of coefficients of the quotient of $P(X)$ after dividing by $X - a$. This is easily proven by an induction on the number of steps involved in dividing $P(X)$ by $X - a$ by long division. In the first step, one tries to divide $a_0 X^n$ by $X - a$. The quotient is $A_0 X^{n-1} = a_0 X^{n-1}$, and to prepare for the next step one subtracts:

$$
\frac{\begin{array}{c} a_0 X^n + a_1 X^{n-1} \\ A_0 X^n - a A_0 X^{n-1} \end{array}}{A_1 X^{n-1}} .
$$

At step $i+1$, one divides $A_{i+1} X^{n-i-1}$ by $X - a$ to get a quotient $A_{i+1} X^{n-i-2}$ and another subtraction (provided $i + 1 \neq n$):

$$
\frac{\begin{array}{c} A_{i+1} X^{n-i-1} + a_{i+2} X^{n-i-2} \\ A_{i+1} X^{n-i-1} - a A_{i+1} X^{n-i-2} \end{array}}{A_{i+2} X^{n-i-2}} .
$$

If one writes $X = Y + a$ and attempts to expand

$$
\begin{aligned}
P(X) = P(Y + a) &= a_0 (Y + a)^n + \ldots + a_{n-1}(Y + a) + a_n \quad (4) \\
&= b_0 Y^n + \ldots + b_{n-1} Y + b_n \\
&= b_0 (X - a)^n + \ldots + b_{n-1}(X - a) + b_n,
\end{aligned}
$$

one can find the b_is by iterating the application of synthetic division. Dividing P by $X - a$ yields the remainder b_n, dividing the quotient yields the remainder b_{n-1}, etc.

There are other methods of determining these coefficients. Occasionally, the Chinese used the Binomial Theorem, as we did in our discussion of the classical construction problems. Newton did the same. Expanding the binomials in (4) and collecting the terms we get:

$$
\left.
\begin{aligned}
b_0 &= a_0 \binom{n}{n} = a_0 \\
b_1 &= a a_0 \binom{n}{n-1} + a_1 \binom{n-1}{n-1} \\
b_2 &= a^2 a_0 \binom{n}{n-2} + a a_1 \binom{n-1}{n-2} + a_2 \binom{n-2}{n-2} \\
&\;\;\vdots \\
b_n &= a^n a_0 \binom{n}{0} + a^{n-1} a_1 \binom{n-1}{0} + \ldots + a_n \binom{0}{0} = P(0).
\end{aligned}
\right\} \quad (5)
$$

Horner's starting point in his 1819 paper was Taylor's Theorem. Writing $X = Y + a = a + Y$, Taylor's Theorem gives

$$P(X) = P(a + Y) = P(a) + \frac{P'(a)}{1!}Y + \ldots + \frac{P^{(n)}(a)}{n!}Y^n$$
$$= P(a) + \frac{P'(a)}{1!}(X - a) + \ldots + \frac{P^{(n)}(a)}{n!}(X - a)^n.$$

For the efficiency of computation, Horner did not calculate all the derivatives but instead worked out the recurrence relations among the b_ks, which essentially amounts to using the formulæ (5) for these coefficients. In his first paper, however, Horner never lost sight of the fact that each b_k for $k < n$ is essentially a derivative. In his later papers, which I have not seen, he is reported to have simplified his account using only algebra.

Getting back to the main point, we can use synthetic division not only to evaluate a polynomial, but also to perform simple substitutions. And this means we can improve approximations to polynomial equations. Being a traditionalist, I will demonstrate the procedure with Qín's oft-cited example (1), starting with the estimate $a = 800$ for a root:

$$P(X) = X^4 - 763200X^2 + 40642560000 \quad \text{and} \quad a = 800. \tag{6}$$

The choice of 800 as a starting point is simply this: the polynomial has a root r with $800 < r < 900$, whence, as an approximation to the root, 800 is correct in the first digit. How Qín knew to start with 800 is not explained in any of the accounts I read. Lǐ and Dù offer no explanation, and Joseph suggests "trial and error". Personally, I think Qín can be credited with having some method. Notice, for example, that the sole negative term is the square one with coefficient of order almost 10^6. It has to cancel a fourth degree term and a constant of order just over 10^{10}, which means that the square of the root should be roughly of order 10^4, and the root itself of order 10^2. This would lead one to evaluating the polynomial by synthetic division at $100, 200, \ldots, 900$, successively. Doing so reveals sign changes between 200 and 300 and between 800 and 900. Of the two revealed positive roots, Qín chose to find the larger.

The successive synthetic divisions proceed as follows:

```
800|1      0  -763200            0   40642560000
        800   640000   -98560000  -78848000000
800|1  800  -123200   -98560000  -38205440000   = b₄
        800  1280000   925440000
800|1 1600  1156800   826880000   = b₃
        800  1920000
800|1 2400  3076800   = b₂
        800
800|1 3200   = b₁

     1    = b₀
```

Thus, for $X = Y + 800$,

$$P(X) = Y^4 + 3200Y^3 + 367800Y^2 + 826880000Y - 38205440000. \qquad (7)$$

The coefficients of (7) are much uglier than those of (6), but since $Y = X - 800$, its root r_1 is $r - 800$, r being the root of (6), and since $800 < r < 900$, we have $0 < r_1 < 100$. That is, the root of (7) is one order of magnitude smaller than that of (6). We can now go on to repeat the process, in this case because we know $0 < r_1 < 100$, we can be a bit more systematic in looking for an approximation to the root r_1 of (7). We can perform synthetic divisions on $Y = 10, 20, 30, \ldots$ until we find a pair yielding opposite signs— or, in this case until we try 40 and discover it to be the root itself and not merely an approximation thereto. And the root of (6) is thus $840 = 800 + 40$. [Although it is irrelevant, I can't help but note that the other root is $X = 240$. It seems clear that Qín concocted the equation from the roots and the question of how he came to consider 800 as his starting point has a trivial solution: he knew it beforehand.[2]]

The unmodified Chinese approach thus proceeds by a succession of stages in determining the decimal expansion of the root one digit at a time, unless the expansion terminates and one will hit upon the exact solution at the corresponding stage. The procedure is slow, but determined, is performed mostly using additions, but also, at each stage, a multiplication by the same essentially single digit number. (Ignoring the 0s, the multiplier in the synthetic division pictured above was always 8.)

Probably the first major improvement in this process was made by Isaac Newton, who didn't really use what we now call Newton's Method.

We are taught in beginning Calculus that Newton's Method for solving an equation

$$f(x) = 0 \qquad (8)$$

works like this. One starts with an initial approximation x_0 to a solution to (8). Then one finds the equation of the tangent line to the curve $y = f(x)$ at the point $(x_0, f(x_0))$:

$$\frac{y - f(x_0)}{x - x_0} = f'(x_0).$$

Finally, one finds where this intersects the x-axis, i.e. one sets $y = 0$ and solves for x:

$$x = x_0 - \frac{f(x_0)}{f'(x_0)}.$$

This point is the next approximation x_1 to a solution to (8), and

$$x_2 = x_1 - \frac{f(x_1)}{f'(x_1)}$$

---

[2] But see Footnote 21, below.

is the next approximation after that. Under quite general conditions, the sequence $x_0, x_1, x_2, \ldots$ converges fairly rapidly to a root of f. This is the procedure as described by Joseph Raphson, not by Newton. However, it is equivalent in execution and has come to be called Newton's Method or the Newton-Raphson Method.

Newton illustrated his method using the equation,

$$X^3 - 2X - 5 = 0. \tag{9}$$

His instruction begins:

> Let the equation $y^3 - 2y - 5 = 0$ be proposed and let the number 2 be found, one way or another, which differs from the required root by less than its tenth part.[3]

With this auspicious beginning, Newton makes the substitution $X = Y + 2$ by appealing to the Binomial Theorem to obtain the equation,

$$Y^3 + 6Y^2 + 10Y - 1 = 0, \tag{10}$$

the solution to which, being the error in using 2 to estimate the solution to (9), is roughly less than $\frac{2}{10}$. Hence, if r is this root, the terms r^3 and $6r^2$ are relatively small and the left hand side of (10) is approximately $10Y - 1$. Thus he solves $10Y - 1 = 0$ and estimates the root of (10) to be .1.

Newton doesn't mention it, but the error in choosing .1 as an estimate for the root of (10) is again less than one-tenth of the actual root. Thus, he is free to repeat the process with the substitution $Y = Z + .1$. The resulting equation is

$$Z^3 + 6.3Z^2 + 11.23Z + .061 = 0.$$

Ignoring the higher degree terms, he estimates the root of this equation to be

$$Z = \frac{-.061}{11.23} = -.0054,$$

after rounding. Setting $Z = W - .0054$ and repeating the process with some rounding produces $W = -.00004852$ as an approximation to the next root. Putting everything together yields

$$X = 2 + .1 - .0054 - .00004852 = 2.09455148$$

as his estimate for the root to (9). This is correct in all 8 exhibited decimals.

Newton's variation on the theme is an advance, but it comes at a price. It gives us more rapid convergence, but we had to multiply by multidigit

---

[3] Newton's presentation of his method is from *De Methodus Fluxionum et Sierierum infinitorum,* which is printed in the third volume of D.T. Whiteside, ed., *The Mathematical Papers of Isaac Newton,* Cambridge University Press, Cambridge, 1969. The relevant excerpt is reprinted in Jean-Luc Chabert's *A History of Algorithms, From the Pebble to the Microchip,* for which cf. the Bibliography.

numbers and even perform some divisions. Moreover, we had to start closer to the root.[4] The issue of convergence now arises. Newton's demand that the error of the initial approximation be no greater than one-tenth the value of the root is neither necessary (cf. the most recent footnote) nor sufficient: the roots of

$$X^2 - 4X + 3.9999 = 0$$

are 1.99 and 2.01. The number 2 is within 10 percent of each of these, but the derivative vanishes at 2 and a second approximation cannot be found via Newton's schema. The derivative does not vanish at 1.9999, which is within 10 percent of each root and Newton's technique produces as the next approximation 1.49995, which is not within 10 percent of either root. Applying Newton's Method to this value results in 1.74987501, again outside the 10 percent limit. Nonetheless, the iteration does converge slowly to the root 1.99. The issue of convergence was first raised by Jean-Reymond Mouraille in 1768 in a paper in which he illustrated ways in which Newton's Method could misbehave, the sequence of approximations converging to the wrong root or not converging at all. The issue was only finally satisfactorily settled by Augustin Louis Cauchy in 1829 in his *Leçons sur le Calcul différential*, wherein he gave quite general conditions under which the procedure was guaranteed success. I refer the reader to Jean-Luc Chabert's *A History of Algorithms* for excerpts of Mouraille's and Cauchy's texts.

Before moving on to Horner, it might be instructive to pause and consider Newton's condition and why it suffices for his example. Let f be a twice-differentiable function with root r. Using Taylor's Theorem with remainder,

$$f(r) = f(x) + f'(x)(r - x) + \frac{f''(\theta)}{2}(r - x)^2,$$

for some θ between r and x. Since $f(r) = 0$, we have

$$0 = f(x) + f'(x)(r - x) + \frac{f''(\theta)}{2}(r - x)^2$$

$$\frac{-f(x)}{f'(x)} = r - x + \frac{f''(\theta)}{2f'(x)}(r - x)^2$$

where we assume $f'(x) \neq 0$. thus

$$x - \frac{f(x)}{f'(x)} = r + \frac{f''(\theta)}{2f'(x)}(r - x)^2.$$

Now, if x_0 is an initial approximation to r, then $x_1 = x_0 - \frac{f(x_0)}{f'(x_0)}$ is the next Newtonian approximation and for all n,

---

[4] Compare this with Qín's treatment. The starting point 800 is within 84 of the root 840, but for the other positive root, 240, Qín's method requires one to start at 200, which is not within 24 of the root.

$$|x_{n+1} - r| = \left| \frac{f''(\theta_n)}{2f'(x_n)} \right| (r - x)^2.$$ (11)

For $f(x) = x^3 - 2x - 5$ we have

$$f'(x) = 3x^2 - 2$$
$$f''(x) = 6x.$$

In the interval $[2, 2.2]$,

$$\left. \begin{array}{l} f'(x) \geq 3 \cdot 2^2 - 2 = 10 \\ f''(x) \leq \quad 6(2.2) \quad = 13.2. \end{array} \right\}$$ (12)

For $x_n \in [2, 2.2]$, we can plug these values into (11) to obtain

$$|x_{n+1} - r| \leq \frac{13.2}{2(10)} (r - x_n)^2$$
$$\leq .66(r - x_n)^2$$ (13)
$$\leq .66 \frac{r^2}{100} = .0066r^2$$
$$< .0066(2.2r)$$
$$< .01452r,$$

and x_{n+1} is again within 10 percent of r. Moreover, since r is close to the centre of the interval, this inequality puts x_{n+1} into $[2, 2.2]$ and the inequalities (12) apply to x_{n+1}.

To estimate the rapidity of convergence, begin with Newton's $x_0 = 2$ and $r < 2.1$:

$$|x_0 - r| < .1$$
$$|x_1 - r| < .66(x_0 - r)^2, \text{ by (13)}$$
$$< .66(.1)^2 = .0066$$
$$|x_2 - r| < .66(x_1 - r)^2$$
$$< .66(.0066)^2 < .00002875$$
$$|x_3 - r| < .66(x_2 - r)^2$$
$$< .00000000055.$$

We see immediately that each step is indeed doubling the number of digits secured for the root.

Perhaps because of the generality of Newton's Method as applied by Joseph Raphson, or because of the geometric presentation as described by Mouraille, the substitutive aspect of Newton's Method did not come to the fore until the early 19th century when Paolo Ruffini, Ferdinand François Désiré Budan, and William George Horner entered the scene.

The first to publish was Ruffini, whose 1804 paper contains a clear presentation of the use of iterated synthetic division to effect a substitution. Budan's initial paper of 1807 showed how to find the coefficients of the polynomial $P(Y + 1)$ from those of $P(X)$ using only additions. In 1813, he published the general result with an algorithm equivalent to Ruffini's.

Horner's paper of 1819 was written in ignorance of Ruffini's, but with knowledge of Budan's 1807 paper. In addition to finding Budan's algorithm insufficiently general, he noted that "its extremely slow operation renders it perfectly nugatory". To demonstrate how his method compared favorably with other methods, he applied it to Newton's equation (9) and in three stages of approximation arrived at the root,

$$2.09455148152326590,$$

correct to the 18th decimal place. "So rapid an advance is to be expected only under very favorable data. Yet this example clearly affixes to the new method, a character of unusual boldness and certainty".

Horner's initial paper of 1819, entitled "A new method of solving numerical equations of all orders, by continuous approximation", appeared in the *Philosophical Transactions* of the Royal Society, having been presented by Davies Gilbert, Esq. FRS. I have a copy of it before me and find it rather opaque. I am not alone in this and can report that Julian Lowell Coolidge, in *The Mathematics of Great Amateurs,* exhibits even greater difficulty with the paper than I had. This might be a good place to quote de Morgan, who evidently understood Horner better. The *Budget*[5] has two informative remarks, the first under the heading "Curious Calculations":

Another instance of computation carried to a paradoxical length, in order to illustrate a method, is the solution of $x^3 - 2x = 5$, the example given of Newton's method, on which all improvements have been tested. In 1831, Fourier's posthumous work on equations showed 33 figures of solution, got with enormous labour. Thinking this a good opportunity to illustrate the superiority of the method of W.G. Horner, not yet known in France, and not much known in England, I proposed to one of my classes, in 1841, to beat Fourier on this point, as a Christmas exercise. I received several answers, agreeing with each other, to 50 places of decimals. In 1848, I repeated the proposal, requesting that 50 places might be exceeded. I obtained answers of $75, 65, 63, 58, 57,$ and 52 places. But one answer, by Mr. W. Harris Johnston, of Dundalk, and of the Excise Office, went to 101 decimal places.

His second remark is a bit more historical in character, and a bit more amusing, and can be found under the heading "Horner's Method":

---

[5] *A Budget of Paradoxes* is cited in full in the Bibliography. My quotations are from pp. 66 - 67 and 187 - 189 of volume II of the second edition.

I think it may be admited that the indisposition to look at and encourage improvements of calculation which once marked the Royal Society
is no longer in existence. But not without severe lessons. They had the
luck to accept Horner's now celebrated paper, containing the method
which is far on the way to become universal: but they refused the
paper in which Horner developed his views of this and other subjects:
it was printed by T.S. Davies after Horner's death. I make myself
responsible for the statement that the Society could not reject this
paper, yet felt unwilling to print it, and suggested that it should be
withdrawn; which was done.

. . .

Horner's method begins to be introduced at Cambridge: it was published in 1820. I remember that when I first went to Cambridge (in
1823) I heard my tutor say, in conversation, there is no doubt that
the true method of solving equations is the one which was published
a few years ago in the *Philosophical Transactions.* I wondered it was
not taught, but presumed that it belonged to the higher mathematics.
This Horner himself had in his head: and in a sense it is true; for all
lower branches belong to the higher.

. . .

It was somewhat more than twenty years after I had thus heard a Cambridge tutor show some sense of the true place of Horner's method,
that a pupil of mine who had passed on to Cambridge was desired
by his college tutor to solve a certain cubic equation— one of an
integer root of two figures. In a minute the work and answer were
presented, by Horner's method. "How!" said the tutor, "this can't be,
you know." "There is the answer, Sir!" said my pupil, greatly amused,
for my pupils learnt, not only Horner's method, but the estimation
it held in Cambridge. "Yes!" said the tutor, "there is the answer certainly; but it *stands to reason* that a cubic equation cannot be solved
in this space." He then sat down, went through a process about ten
times as long, and then said with triumph: "There! that is the way to
solve a cubic equation!"

Coolidge's confused account of Horner in *The Mathematics of Great Amateurs* announces four papers of Horner:

1. "On the popular method of approximation", *Leybourn's Repository,* NS
   IV, 1819

2. "A new method of solving numerical equations", *Ladies' Diary,*[6]1838

3. "On algebraic transformations", *The Mathematician,* I, 1843 - 1856

4. "Horæ arithmeticæ", *Leybourn's Repository,* NS V, 1830.

---

[6] *Lady's Diary* according to Coolidge.

He does not explain what the first of these is and most reference works do not list it. The second is a reprinting of the paper in the *Philosophical Transactions*. According to Coolidge, the third was submitted to the Royal Society in 1823, "but was inserted in *The Mathematician* for 1845, and finally published in 1855". I have seen none of these papers, but the third and fourth are reputedly completely algebraic in character, presumably setting out Horner's Method in a form recognisable today. It is the unrecognisable and all but unreadable *Transactions* paper, reprinted in *Ladies' Diary*, that merits consideration.

Coolidge bases his account on the version in *Ladies' Diary*, although the original *Transactions* paper ought to have been equally accessible to him, and, if not, the nearly complete reprint of it in Smith's source book definitely was. I stress this point because he seems to be describing the original paper, but yet he isn't. There are two explanations for this. Either the reprint was not exact, or Coolidge based his account in those pre-xerox days on notes he had earlier taken from the paper.

The first example Horner gives of the working of his method is to a polynomial that had been considered by Euler:

$$P(X) = X^4 - 4X^3 + 8X^2 - 16X + 20 = 0. \qquad (14)$$

He offers the table

$x =$	0	1	2
	20	9	4
	−16	−8	0
	8	2	8
	−4	0	4
	1	1	1

The columns below the first row represent the coefficients of the polynomials $P(X), P(X+1), P(X+2)$, respectively, in ascending order. They were not found by synthetic division, but by the application of the binomial coefficients:

$$9 = 1(20) + 1(-16) + 1(8) + 1(-4) + 1(1)$$
$$-8 = 1(-16) + 2(8) + 3(-4) + 4(1)$$
$$2 = 1(8) + 3(-4) + 6(1)$$
$$0 = 1(-4) + 4(1)$$
$$1 = 1(1)$$
$$4 = 1(9) + 1(-8) + 1(2) + 1(0) + 1(1)$$
$$0 = 1(-8) + 2(2) + 3(0) + 4(1)$$

etc.

He says

> We need proceed no farther, for the sequences $2, 0, 1$ in the second column, and $4, 0, 8$ in the third, show that the equation has two pairs of imaginary roots. Consequently it has no real roots.

Coolidge says

> In the first column are no permanences of sign, hence the equation
> has no negative roots. In the third column if we treat 0 as positive the
> reduced equation has no positive roots; hence the original one had no
> positive roots greater than 2. He concludes that as the once reduced
> equation is close to $2x^2 + 1 = 0$ it has no real roots.

These don't seem to be the same reasons.

In fact, Coolidge has it wrong. One page before presenting this example, Horner reminds the reader of a result of Jean Paul de Gua de Malves by which, given a polynomial $Q(X)$ with real coefficients, if a number a exists for which $Q^{(k)}(a)$ and $Q^{(k+2)}(a)$ have the same nonzero sign and $Q^{(k+1)}(a) = 0$ then Q has a pair of complex roots. Moreover, Horner says, if this happens for distinct a, b, distinct pairs of such roots exist. The coefficients of a polynomial being essentially the successive derivatives of the polynomial, this means that a coefficient of 0 between two nonzero coefficients of like sign— as $2, 0, 1$ or $4, 0, 8$— signifies a pair of complex roots. And two such triples in different columns signify distinct such pairs of complex roots. The quotation from Horner makes clear that de Gua's result is the reason he says Euler's equation (14) has 4 complex and hence no real roots.

Coolidge's rejection of real roots depends initially on Descartes' Rule of Signs, or rather, a trivial case of it: if all the coefficients of a polynomial with nonzero constant term are nonnegative, then the polynomial has no positive roots. Thus, for Euler's $P(X)$, $P(X + 2)$ satisfies this condition and has no positive roots— whence $P(X)$ has no roots greater than 2. Because the coefficients of $P(X)$ alternate in sign, $P(-X)$ has no positive roots, i.e. $P(X)$ has no negative roots. Thus any root of $P(X)$ must lie between 0 and 2. Now $P(X + 1)$ is $X^4 + 2X^2 - 8X + 9$, which is not at all close to $2X^2 + 1$ in that interval and I have no idea why Coolidge should think it is.[7]

There are two fairly obvious explanations why Coolidge gives a different reason from Horner's for concluding all four roots to be complex. One is that the reprint of Horner's paper in the *Ladies' Diary* is an amended one and Horner himself, or his editor[8], has given a new reason. The other is that Coolidge was reproducing the paper from his handwritten notes and, not having the reference to de Gua at hand, filled in the missing steps as best he

---

[7] One can, however, get a correct de Gua-free proof by observing that

$$\frac{d}{dX}P(X + 1) = 4X^3 + 4X - 8 < 0 \quad \text{for} \quad -1 < X < 1.$$

Hence $P(X+1)$ decreases monotonically from 20 to 4 on $(-1, 1)$ and $P(X)$ has no root between 0 and 2. Or, if one is wedded to $2X^2 + 1$, subtract it from $P(X + 1)$ to get $X^4 - 8X + 8$. Both X^4 and $8 - 8X$ are nonnegative in $(-1, 1)$, whence $P(X + 1) \geq 2X^2 + 1 > 0$ in that interval.

[8] Horner died in 1837, a year before the new appearance of his paper.

could. His book was published 9 years after his retirement and, he being in his mid-70s, his mathematics may have been a little rusty.[9]

Is this an important enough point to check Horner's paper in the *Ladies' Diary*? The *Transactions* paper is readily available. It is reproduced almost in full in Smith's *Source Book,* and, although my own college is too new to have the original journal, the larger universities probably have it and, in any event, my college does subscribe to *JSTOR* and I was able to download a copy of the original in *PDF* format. The *Ladies' Diary* is a different story.

The *Ladies' Diary; or, Complete Almanack,* sometimes called the *Ladies' Diary; or, Woman's Almanack,* began publication in 1704. Around 1788 and for a period the duration of which I do not know, it had *The Diary Companion, Being a Supplement to the Ladies' Diary.* After 1840 it merged with the younger *Gentleman's Diary, or the Mathematical Repository, an Almanack* (b. 1741) to become the *Lady's and Gentleman's Diary,* which lasted until 1871. Journal mottoes over time include

> Designed principally for the amusement and instruction of the fair sex;

and

> Containing new improvements in arts and sciences, and many interesting particulars: designed for the use and diversion of the fair sex;

while the gentlemen read

> Many useful and entertaining particulars, peculiarly adapted to the ingenious gentleman engaged in the delightful study and practice of mathematics;

and ladies and gentlemen shared a common diary

> Designed principally for the amusement and instruction of students in mathematics: comprising many usefully and entertaining particulars, interesting to all persons engaged in that delightful pursuit.

Whereas the *Philosophical Transactions* of the Royal Society is the official organ of that august body of British savants, the *Ladies' Diary* was written at a more popular level, but at a much higher level than would be popular today. Educational opportunities for young women were limited and the *Ladies' Diary* was designed to help fill the gap. Its readership would certainly have had the algebraic and computational background required by Horner's paper. But what about Calculus? Well, there isn't that much involved and one can read Horner's paper knowing only that the derivatives of a given polynomial are positive multiples of the coefficients of the polynomial. In any event, the

---

[9] Indeed, he added a section on Qín's equation (1) at the end of his chapter on Horner and remarked that "if we solve this as a quadratic in x^2 we find four real approximate roots $x = \pm 76, x = \pm 265$. But Ch'in, for some reason I do not grasp, takes as his first approximation 800"!!

Calculus doesn't enter into Horner's treatment of Euler's equation, except possibly in the proof [10] of de Gua's cited result. However, Horner does appeal to the Calculus again in his second example and Coolidge repeats this. So I think it is safe to conclude that Horner's exposition would not have been alterred on account of the intended audience.[11]

This leaves the possibility that a change was made to correct an error. In looking for a probable error, the only place my finger wants to point to is the application of de Gua's result. This result, as Horner attributes it to de Gua, is an easy corollary to Descartes' Rule of Signs; Horner's generalisation is not so trivial, but it is certainly correct. In short, I see no mathematical necessity of substituting a different treatment, certainly not an incorrect one, for Euler's equation.

All in all, I am more inclined to believe that Coolidge erred than that this part of Horner's paper was rewritten. Consulting the *Ladies' Diary* does not seem to be necessary.[12]

As his second example, Horner cites an equation studied by Lagrange. The problem is "to determine the nearest distinct limits of the positive roots" of

$$X^3 - 7X + 7 = 0. \tag{15}$$

This equation has one negative and two positive roots and we are supposed to approximate the positive roots. Starting at 0, Horner substitutes $a = 1, a = 2$, stopping at this latter because all the coefficients for $P(X + 2)$ are positive:

$x =$	0	1	2
	7	1	1
	−7	−4	5
	0	3	6
	1	1	1

Looking at the derivatives of P (third row), he concludes $P'(x) = 0$ for some $1 < x < 2$, and that the roots, if they exist, lie on either side of this x. As the second derivatives are positive at 1 and 2 (fourth row), this suggests the second derivative at x will also be positive and thus P will assume a minimum

---

[10] Cf. the section on de Gua's Theorem later in this chapter.

[11] More information on the *Ladies' Diary* and the high level of sophistication of its audience, as well as the composition thereof, can be found in Teri Perl, "The Ladies' Diary or Woman's Almanack, 1704 - 1841", *Historia Mathematica* 6, (1979), pp. 36 - 53.

[12] And a good thing! Unlike the *Philosophical Transactions,* the *Ladies' Diary* is somewhat rare. A perfunctory search on the Internet revealed two universities in my area that have the 1838 volume in their rare book rooms, which probably explains why I couldn't acquire the article via Interlibrary Loan. The one university is a two or three hour drive from me, and the other is a private school with restricted access to its library, which means preliminary work acquiring permission before taking the one hour drive to get there.

at x. This tells us nothing definite, but suggests we take a closer look in the interval $[1, 2]$. Horner subdivides the interval and constructs the table:

$X =$	1.0	1.1	1.2	1.3	1.4	1.5	1.6	1.7
	1000	631	328	97	−56	−125	−104	13
	−400	−337	−268					
	30	33	36					
	1	1	1					

and concludes the roots to lie between 1.3 and 1.4 and between 1.6 and 1.7.

Some explanation of this new diagram is in order. Coolidge says ,"When he wishes to avoid decimals he multiplies all roots by 10". Actually, he has performed another substitution $10X = Y$ on $P_1(X) = X^3 + 3X^2 - 4X + 1$ to get

$$Q(Y) = \left(\frac{Y}{10}\right)^3 + 3\left(\frac{Y}{10}\right)^2 - 4\left(\frac{Y}{10}\right) + 1,$$

and

$$1000\, Q(Y) = Y^3 + 30Y^2 - 400Y + 1000.$$

And the first row of the table should read

$$Y = \quad 10 \mid \quad 11 \quad 12 \quad 13 \quad 14 \quad 15 \quad 16 \quad 17.$$

As we move from left to right he fills down the columns using the formulæ (5), with the differences in argument at each step being 1— eliminating the multiplications by a in these formulæ. In particular, the last formula reads

$$b_3 = a_0 \binom{3}{0} + a_1 \binom{2}{0} + a_2 \binom{1}{0} + a_0 \binom{0}{0} = a_0 + a_1 + a_2 + a_3,$$

and $631 = 1 + 30 - 400 + 1000$. For the third row, we have

$$b_2 = a_0 \binom{3}{1} + a_1 \binom{2}{1} + a_2 \binom{1}{1} = 3a_0 + 2a_1 + a_2,$$

and $-337 = 3 + 2 \cdot 30 - 400$. Etc. When the second column is filled, he repeats the process to obtain the third column. The first element, 97, of the fourth column is obtained the same way. After that there is no longer any need to determine full columns: Horner used finite difference methods to extend the series $1000, 631, 328, 97, \ldots$ I will not go into these, other than to mention that the Chinese had been familiar with finite differences and used them centuries earlier, either about the time Qín was explaining "Horner's method" or shortly thereafter.

Horner next turns to explaining the workings of his algorithm in the cases of equations of low degree. In the linear case, it reduces to long division, and in the quadratic case to the "known arithmetical process for extracting the square root". The workings in the cubic case are totally unfamiliar and with his opaque notation the schematic diagram of the procedure is no help to the casual reader like myself or, apparently, Coolidge. Horner presents an example of a simple cube root extraction. He then applies the procedure to Newton's equation (9).

He begins by observing that the solution to $X^3 - 2X - 5 = 0$ is a little greater than 2. Hence the substitution $Y = X + 2$, resulting in the equation,

$$Y^3 + 6Y^2 + 10Y - 1 = 0.$$

À la Newton, he knows the next approximation must be close to the root .1 of

$$10Y - 1 = 0.$$

But he refines this, figuring Y is also close to a solution to

$$6Y^2 + 10Y - 1 = 0$$
$$(6Y + 10)Y - 1 = 0$$
$$(6(.1) + 10)Y - 1 = 0$$

since Y is close to .1 and he thus solves this last,

$$Y = \frac{1}{10.6} \approx .094,$$

to use .094 as his corrected estimate. Under the familiar Horner schema, this would mean looking at $Z = Y + .094$:

$$Z^3 + 6.2822Z^2 + 11.154508Z - .006153416.$$

Ignoring the first two terms gives

$$Z \approx .0005.$$

Then

$$6.282(.0005) + 11.154508 \approx 11.1576491 \approx 11.158,$$

and he takes as his correction

$$\frac{.006153416}{11.158} \approx .0005314801936 \approx .00055148.$$

At the next step he gets the correction term,

$$.000000001542326590,$$

and declares the solution to Newton's equation to be

$$2.094551481542326590,$$

correct to 18 decimals in three steps.

Coolidge comments on this:

He goes at some length into the equation

$$x^3 - 2x - 5 = 0.$$

There is clearly a root a little greater than 2. Reducing by that amount,

$$x^3 + 6x^2 + 10x - 1 = 0.$$

This has a root slightly less than .1. He tries 0.09. He then covers up most of his work, finally coming out with the answer

$$x = 2.094551481542326590,$$

'correct to the 18th decimal place at three approximations.'

Actually, Horner says "correct *in* the 18th decimal place...", and he tries .094, not .09. Coolidge's .09 results in 2.09, correct to two decimal places. The next estimate 2.0945513656 [13] is correct to only 6 decimal places and not the 8 places two iterations gave Horner. One can only expect 14 places after 3 iterations... Moreover, Horner, although not as explicit as I have been in the working of his method, did not in any way "cover up" most of his work. The method of correcting his estimates is not unlike the correction used in finding square roots and his intended readers would have had no trouble with it.

Coolidge distrusts Horner's claim to accuracy, citing the following example:

$$X^3 + 25X^2 + 5X - .961725 = 0.$$

The first digit of the solution is approximately .1, whence we make the substitution $X = Y + .1$ and obtain

$$Y^3 + 25.3Y^2 + 10.03Y - .210725 = 0.$$

This gives $Y = .02100094716$, which we can truncate to .02. He concludes that the next figure is .02, yielding .12 which should be correct to two decimal places but isn't. The error, however, is less than 10^{-6} and is a better estimate than .01, which is correct to 2 decimals. However, Coolidge neglected the correction,

$$Y = \frac{.210725}{25.3(.02) + 10.3} \approx .0195007403 \approx .019,$$

which is correct to 3 decimals.[14] Another iteration yields

$$X \approx .1199997$$

correct to 7 decimals.

---

[13] This is the value that arises by truncating the initial linear estimate to 4 decimal places. Allowing more still yields only 6 correct digits in the corrected estimate.

[14] I have rounded down.

Coolidge's attempted counterexample brings to mind the Cambridge tutor's objection that "it stands to reason"... As I quoted earlier, Horner himself said of his application to Newton's equation, "so rapid an advance is to be expected only under very favorable data", which Newton's example supplies. That said, such rapid convergence is nothing to be surprised at. We saw it above with Newton's Method and it should be expected here. A little Calculus bears this out. Horner's method starts with a polynomial $P(X)$ and an approximation r to a root of P, and then proceeds to consider the substituted polynomial

$$P_r(Y) = P(Y + r) = P(r) + P'(r)Y + \frac{P''(r)}{2}Y^2 + \frac{P'''(r)}{6}Y^3 + \ldots, \quad (16)$$

and tentatively choosing

$$Y = \frac{-P(r)}{P'(r)}$$

as the approximation to a root of (16), i.e.

$$X = Y + r = r - \frac{P(r)}{P'(r)}$$

is the tentative next approximation to a root of P. Now, this is just Newton's approximant and was chosen because, r being close to the root, Y will be very small and the higher order terms of (16) will not come very strongly into play. Horner's corrected estimate is obtained by now combining the Newtonian estimate and an additional term by solving for Y in the approximation to (16):

$$0 = P(r) + P'(r)Y + \frac{P''(r)}{2}\left(\frac{-P(r)}{P'(r)}\right)Y.$$

Horner's corrected estimate to the solution to (16) is thus

$$Y = \frac{-P(r)}{P'(r) - \frac{P(r)P''(r)}{2P'(r)}} = \frac{-2P(r)P'(r)}{2(P'(r))^2 - P(r)P''(r)},$$

whence

$$X = r - \frac{2P(r)P'(r)}{2(P'(r))^2 - P(r)P''(r)}$$

is his next estimate to a root of P.

Thus, define

$$H(X) = X - \frac{2P(X)P'(X)}{2(P'(X))^2 - P(X)P''(X)}$$

$$= \frac{2XP'(X)^2 - XP(X)P''(X) - 2P(X)P'(X)}{2P'(X)^2 - P(X)P''(X)}. \quad (17)$$

Theoretically, if not computationally, Horner's method consists of calculating the sequence $x_0 = r$, $x_1 = H(x_0)$, $x_2 = H(x_1)$, ... and watching the x_n's converge to a root, say, ρ. The Mean Value Theorem allows us to estimate the rapidity of convergence[15]:

$$|x_{n+1} - \rho| = |H(x_n) - H(\rho)| = |x_n - \rho| \cdot |H'(\theta)| \qquad (18)$$

for some θ between ρ and x_n. The numerator of $H'(X)$ is

$$\Big[\Big(2P'(X)^2 + 4XP'(X)P''(X) - P(X)P''(X) - XP'(X)P''(X)$$
$$- XP(X)P'''(X) - 2P'(X)^2 - 2P(X)P''(X)\Big)$$
$$\Big(2P'(X)^2 - P(X)P''(X)\Big)$$
$$- \Big(2XP'(X)^2 - XP(X)P''(X) - 2P(X)P'(X)\Big)$$
$$\Big(4P'(X)P''(X) - P'(X)P''(X) - P(X)P'''(X)\Big)\Big]$$

which is congruent modulo $P(X)$ to

$$3XP'(X)P''(X)\cdot 2P'(X)^2 - 2XP'(X)^2 \cdot 3P'(X)P''(X)$$
$$\equiv 6XP'(X)^3 P''(X) - 6XP'(X)^3 P''(X)$$
$$\equiv 0 \quad \bmod P(X).$$

Thus, $P(X)$ is a factor of the numerator of $H'(X)$.

The denominators of H and H' are 0 at only finitely many points, so, as long as P and P' do not share ρ as a root, $H'(x)$ will be very close to 0 when x is close to ρ. This means convergence will be rapid. For Newton's equation, starting at $x_0 = 2$, the error is less than .1, and we can calculate[16]

$$H'(2) = .0069419722 < 10^{-2},$$

whence, for $x_1 = 2.094$,

$$|x_1 - \rho| < |x_0 - \rho| \cdot 10^{-2} < 10^{-1} \cdot 10^{-2} = 10^{-3}.$$

For the next approximation, $x_2 = 2.09455148$, we have[17]

---

[15] Here, I assume convergence and merely address the issue of rapidity of convergence.

[16] On the interval $[2, 2.1]$, H' decreases to a minimum value of 0 at ρ. As we have $x_0 < x_1 < \ldots$, $H'(x_n)$ can be used as an upper bound for $H'(\theta)$ in (18).

[17] It may be worth noting that if we follow Coolidge's account of Horner and take $x_1 = 2.09$, we will only get $H'(2.09) = .00001421745544 < 10^{-4}$. Since $|2.09 - \rho| < 10^{-2}$, this will yield an estimate within 10^{-6}, i.e. correct to 6 decimal places as in Coolidge's analysis.

$$H'(2.094) = .00000020759 < 10^{-6}$$

whence

$$|x_2 - \rho| < 10^{-3} \cdot 10^{-6} = 10^{-9}.$$

At this point my trusty *TI-83+* can no longer handle the numbers properly. Writing $H(X) = Q(X)/R(X)$, if I enter an expression for

$$\frac{Q'(X)R(X) - Q(X)R'(X)}{R(X)^2} \tag{19}$$

into the calculator and evaluate it at 2.09455148 I get 0. If I multiply (19) out to get

$$H'(X) = \frac{18X^8 - 66X^6 - 180X^5 + 48X^4 + 300X^3 + 474X^2 + 120X + 150}{(6X^4 - 6X^2 + 15X + 4)^2}$$

and evaluate this, the calculator tells me

$$H'(2.09455148) = -1.9330394 \times 10^{-14},$$

and if I factor the numerator

$$H'(X) = \frac{6(X^3 - 2X - 5)^2(3X^2 + 1)}{(6X^4 - 6X^2 + 15X + 4)^2},$$

and perform the evaluation, I find

$$H'(2.09455148) = 1.62245219 \times 10^{-18}.$$

All three values are incorrect. The mild extra precision of the *TI-85* gives

$$H'(2.09455148) = 1.62245219229 \times 10^{-18}$$

for the factored form, but this is still incorrect. [I might also note that, for the expanded form, the *TI-85* gives $-1.93303935019 \times 10^{-14}$, agreeing with the *TI-83+* as above.]

One can, of course, stop at this point and say that it doesn't matter; the point has been made: there is nothing suspicious about Horner's claims to great accuracy; Coolidge simply did not understand what Horner was doing. That is the easy way out and is thus to be rejected out of hand. One could program one's calculator to handle numbers with higher precision, e.g. by using lists. Or, one could set aside the calculator and actually analyse the situation. Notice that for any x,

$$\begin{aligned}
x^3 - 2x - 5 &= (x^3 - 2x - 5) - (\rho^3 - 2\rho - 5) \\
&= (x^3 - \rho^3) - 2(x - \rho) \\
&= (x - \rho)(x^2 + x\rho + \rho^2 - 2),
\end{aligned}$$

whence on $[2, 2.1]$,

$$|x^3 - 2x - 5| \le |x - \rho| \cdot (3(2.1)^2 - 2) \le 11.23|x - \rho|.$$

Thus,

$$|H'(x)| \le \frac{6(11.23)^2|x - \rho|^2(3(2.1)^2 + 1)}{(6x^4 - 6x^2 + 15x + 4)^2}$$

$$< \frac{10770|x - \rho|^2}{(6x^4 - 6x^2 + 15x + 4)^2} \le \frac{10770|x - \rho|^2}{106^2} \tag{20}$$

$$< |x - \rho|^2, \tag{21}$$

where (20) follows from the fact that $6X^4 - 6X^2 + 15X + 4$ is monotone increasing on $[2, 2.1]$[18] and evaluates to 106 at 2. Now, combining (18) and (21) we have

$$|x_{n+1} - \rho| \le |x_n - \rho| \cdot |H'(x_n)| < |x_n - \rho|^3.$$

Starting with $x_0 = 2$ where $|x_0 - \rho| < 10^{-1}$, we get successively

$$|x_0 - \rho| < 10^{-1}$$
$$|x_1 - \rho| < 10^{-3}$$
$$|x_2 - \rho| < 10^{-9}$$
$$|x_3 - \rho| < 10^{-27}$$

and it is the modesty of Horner's 18 decimal places where he should have got 26 that merits investigation!

2 Descartes' Rule of Signs

One way to get the students involved in historical research right off the bat might be to give them, say, a week to answer the following questions:

What is Descartes' Rule of Signs?

What is Descartes' statement of the Rule of Signs?

Who proved Descartes' Rule of Signs?

The point is that Descartes' Rule of Signs has a somewhat complex and even confusing history for such a simple matter. The textbooks do not go into it at all and, assuming conscientious students, when one asks them the results of their researches the following week, they will likely have conflicting reports. For, the secondary references give varying accounts and although the more

---

[18] The second derivative is positive, whence the derivative is increasing. But this derivative is positive at $x = 2$.

complete accounts will have significant overlap, they need not agree on their answers to any of the three questions.

For our purposes, Descartes' Rule is a tool in understanding Horner's application of his method and is not an end in itself. Thus, although I shall make a few historical remarks on its development, I shall not attempt any completeness of coverage.[19] I wish merely to discuss its proof, its application to a proof of de Gua's Theorem, and its use in applying Horner's Method.

The first thing we must do is give a statement of the Rule— or, Rules: there are several variants of the Rule of varying levels of strength and refinement. All versions, however, require some technical jargon.

2.1 Definitions. *Let* $a_0, \ldots, a_n$ *be a sequence of real numbers,* $a_0 \neq 0$. *The sign of* a_i *is 1 if* a_i *is positive,* -1 *if* a_i *is negative, and the sign of* a_{i-1} *if* a_i *is 0. We say that the sequence* $a_0, \ldots, a_n$ *exhibits a* change of signs *between* a_i *and* a_{i+1} *if the signs of* a_i *and* a_{i+1} *are different; it exhibits a* permanence *of signs between* a_i *and* a_{i+1} *if the signs of* a_i *and* a_{i+1} *are the same.*

In this definition we are interested in the actual crossovers from positive to negative or from negative to positive in going from beginning to end of the sequence. Thus, when we hit a 0, although it stands on the border between positive and negative, it does not represent a crossover and we give it the sign of its predecessor. Some prefer to simply delete the 0s from the sequence in determining changes and permanences of sign.

We may for simplicity's sake say that the sequence *changes sign* between a_i and a_{i+1} or in going or passing from a_i to a_{i+1} instead of using the more long-winded "exhibits a change of sign between a_i and a_{i+1}". Similarly, we may refer briefly to a *sign change* for a change of signs and to a *permanence* for a permanence of signs. Some authors refer to *variations* instead of *changes* and/or to *permanencies* instead of *permanences*.

Descartes's Rule refers to the changes and permanences of signs in the sequence of coefficients of a polynomial. As the phrase "sign changes of the sequence of coefficients of a polynomial" is rather long and cumbersome, I prefer "sign changes of a polynomial". This conflicts with common sense, whereby a sign change of a polynomial refers to the change of sign of the value of the polynomial in an interval. However, we can keep the notions separate by agreeing that the passive voice ("sign change of a polynomial") refers to the sequence of coefficients, and the active voice ("a polynomial changes sign") refers to the behaviour of the polynomial.

With all of this, we may state Descartes' Rule of Signs:

2.2 Theorem. *Let* $P(X)$ *be a polynomial with real coefficients.*
i. the number of positive real roots of P (counting multiplicities) is at most

---

The reader who is unafraid of the Italian language is referred to M. Bartolozzi and R. Franci, "La Regola di Segni dall'Enunciato di R. Descartes (1637) alla Dimostrazione di C.F. Gauss (1828)", *Archive for History of Exact Sciences* 45, no. 4 (1993), pp. 335 - 374.

the number of sign changes of P;
ii. the number of negative real roots of P (counting multiplicities) is at most the number of permanences of P.

Theorem 2.2 is Descartes' actual formulation of his unnamed rule, which he stated explicitly in *La Gèomètrie* in 1637. In this he was at least partially anticipated by Thomas Harriot and, to a lesser extent, by Cardano. None of the three offered a hint of a proof. However, Descartes also mentioned the following sharper formulation:

2.3 Theorem. *Let $P(X)$ be a polynomial with real coefficients.*
i. the number of positive real roots of P is at most the number of sign changes of P;
ii. the number of negative real roots of P is at most the number of sign changes of the polynomial $P(-X)$.

In stating this, I have dropped the phrase "counting multiplicities". Let us agree that it is understood.

Descartes did not actually write "$P(-X)$", but described its coefficients and explained that its positive roots were the negatives of the negative roots of P, whence Theorem 2.3.ii is a consequence of Theorem 2.3.i. He did not mention at all that the second formulation is in any way superior to the first. In 1828, Gauss said this was the proper formulation of the Rule because of its greater simplicity. The proof is easier, the statement is more uniform, and, should 0 be a root of P, the multiplicity of 0 as a root has to be subtracted from the estimate given by 2.2.ii. What Gauss did not mention (and we will see a bit later) is that even after adjusting for the multiplicity of 0 as a root, 2.2.ii may still not give as good an estimate as 2.3.ii.

The most popular version today of Descartes' Rule of Signs reads as follows:

2.4 Theorem. *Let $P(X)$ be a polynomial with real coefficients.*
i. the number of positive real roots of P either equals the number of sign changes of P or falls short of this latter number by an even integer;
ii. the number of negative real roots of P either equals the number of sign changes of $P(-X)$ or falls short of this latter number by an even integer.

Finally, I should offer the seldom stated formulation:

2.5 Theorem. *Let $P(X)$ be a polynomial with real coefficients.*
i. the number of positive real roots of P either equals the number of sign changes of P or falls short of this latter number by an even integer;
ii. the number of negative real roots of P either equals the number of permanences of P minus the multiplicity of 0 as a root, or falls short of this difference by an even integer.

Most expositions present Theorem 2.4 as Descartes' Rule of Signs and leave it uncredited, suggesting implicitly that this is Descartes' own formulation. Where credit is assigned, it is to Carl Friedrich Gauss. The result, however, is not stated explicitly by Gauss in his oft-cited 1828 paper[20] on the subject. I would suppose Theorem 2.5 may also be safely attributed to him.[21]

As regards proof, a number of mathematicians attempted such. In my home library, I find the names Gottfried Wilhelm Leibniz, Jean Prestet, Johann Andreas Segner, Jean-Paul de Gua de Malves, Jean Baptiste Joseph Fourier, and Gauss spread over several references. Isaac Newton is cited for having mentioned the result without proof and giving his own more precise method of determining the number of positive roots— a method first rigorously justified by J.J. Sylvester in 1865. De Gua is recognised by all as having published the first adequate proof of Theorem 2.2 in a book in 1740, just over a century after Descartes asserted it as a fact. In one source de Gua's proof is described as geometric, while in another it is stated that, since Segner (who is credited in the *Dictionary of Scientific Biography* with a correct proof in the case in which all roots are real) and de Gua, proofs of Theorem 2.2.i generally proceed by induction, showing for $r > 0$ the polynomial $(X - r)P(X)$ to have at least one more sign change than P.[22] Another source praises Fourier for having found an inductive proof, which he did while yet a teenager. To cite the *Dictionary of Scientific Biography,* "The details of the proof may be seen in any textbook dealing with the rule, for Fourier's youthful achievement quickly became the standard proof, even if its authorship appears to be virtually unknown". The story generally ends with Gauss' elementary proof published in 1828, or it continues in a different direction citing Fourier, Ferdinand François Budan, and Charles François Sturm in counting the number of roots of a polynomial in an interval $[a, b]$.

These histories do not go on to cover generalisations, but there are such: Edmond Nicolas Laguerre allowed for expressions with non-integral exponents in a paper of 1883, and Georg Pólya considered some generalisations in a paper of 1914 and later included a chapter on the subject in his classic *Aufgabe und Lehrsätze aus der Analysis,* co-authored with Gabor Szegö.

---

[20] C.F. Gauss, "Beweis eines algebraischen Lehrsatzes", *Crelle's Journal für die reine und angewandte Mathematik* 3, no. 1 (1828), pp. 1 - 4. An English translation by Stewart A. Levin can be found online.

[21] Lǐ and Dù's *Chinese Mathematics; A Concise History* informs us that Lǐ Rui (1773 - 1817) (rendered Li Juan in Mikami's *Mathematics in China and Japan*) independently discovered Descartes' Rule, albeit after Descartes and Harriot. However, he does take priority in stating, if not in full generality, that the difference between the number of sign changes and the number of roots is even. Mikami credits him as the first Chinese mathematician to be aware of the fact that an equation could have more than one root. Lǐ and Dù credit Wāng Lái with this.

[22] Bartolozzi and Franci, *op.cit.,* say that de Gua provided two proofs, one algebraic and one analytic. They give 1741 as the publication date; the *Dictionary of Scientific Biography* gives 1740.

Generalisations and refinements aside, the job of proving Descartes' Rule of Signs did not end with Gauss. In 1943, the American mathematician A.A. Albert, displeased with the complicated proof in L.E. Dickson's *First Course in the Theory of Equations*, published another proof more complicated than that of Gauss.[23] In 1963, the Indian mathematician P.V. Krishnaiah published a very nice, simple proof[24] of Theorem 2.2.i by performing the induction in the above cited passage from $P(X)$ to $(X - r)P(X)$ on the number of sign changes in the polynomial P instead of on the number of its positive real roots as would seem the more obvious thing to do.

The simplest proof of Theorem 2.4 uses Calculus.

2.6 Lemma. *Let* $P(X) = a_0 X^n + \ldots + a_n, a_0 > 0$.
i. if $a_n > 0$, *the number of positive roots of* P *is even;*
ii. if $a_n < 0$, *the number of positive roots of* P *is odd.*

Proof. Because $a_0 > 0$, P is positive for all sufficiently large arguments and we can choose b so that $P(b) > 0$ and all positive roots of P lie in the interval $[0, b]$.

i. $P(0) = a_n > 0$. Since P starts out positive and finishes up positive in $[0, b]$, P changes sign an even number of times going from 0 to b. [To be ultra-explicit: P changes sign only finitely many times because P must pass through 0 (by the Intermediate Value Theorem) when changing signs, and P has at most n roots.]

Now the roots of P are either of odd order (including order 1 for a simple root) or of even order. Let r be a positive root of order k, and let ϵ be very small. By Taylor's Theorem with Lagrange's form for the remainder,

$$P(r + \epsilon) = P(r) + \epsilon \frac{P'(r)}{1!} + \ldots + \epsilon^{k-1} \frac{P^{(k-1)}(r)}{(k-1)!} + \epsilon^k \frac{P^{(k)}(\theta)}{k!} = \epsilon^k \frac{P^{(k)}(\theta)}{k!},$$

where θ is between r and $r + \epsilon$. Choosing ϵ sufficiently small, $P^{(k)}(\theta)$ can be taken to be nonzero and of the same sign as $P^{(k)}(r)$. Thus, if k is even, $P(r + \epsilon)$ and $P(r - \epsilon)$ are nonzero and have the same sign as $P^{(k)}(r)$; while, if k is odd, $P(r + \epsilon)$ and $P(r - \epsilon)$ are nonzero and have opposite signs.

It follows that at roots of odd order P changes sign, and at roots of even order P does not. Thus the number of times P changes sign in $[0, b]$ equals the number of distinct positive roots of odd order P has. But we've already remarked that the first of these numbers is even, whence P has an even number of distinct positive roots of odd order. Counting multiplicities and adding the multiplicities of the positive roots of even order merely adds an even number to this.

ii. The proof is similar.   □

---

[23] A.A. Albert, "An inductive proof of Descartes' Rule of Signs", *American Mathematical Monthly* 50, no. 3 (1943), pp. 178 - 180. To be fair, I should mention that Albert proved Theorem 2.4 and not just Theorem 2.2.

[24] P.V. Krishnaiah, "A simple proof of Descartes' Rule of Signs", *Mathematics Magazine* 36, no. 3 (1963), p. 190.

2.7 Lemma. *Let $P(X)$ be given. The difference between the number of positive roots of P and the number of sign changes of P is even.*

Proof. We may assume $P(0) \neq 0$. For, if 0 is a root of P of multiplicity k, the sequence of coefficients of $P(X)/X^k$ is just the truncated sequence of coefficients of P with all the end 0s removed— and there is no change in either the numbers of positive roots or sign changes in switching polynomials.

Multiplying P by -1 if necessary, we may also assume the leading coefficient of P to be positive.

Write $P(X) = a_0 X^n + \ldots + a_n$. If $P(0) = a_n > 0$, the number of sign changes is even since the overall transition is from positive to positive. But by Lemma 2.6, P has an even number of of positive roots in this case.

If $P(0) = a_n < 0$, there is an odd number of sign changes and, by Lemma 2.6, P has an odd number of positive roots. The difference is even.   □

2.8 Lemma. *Let the polynomial P have m positive roots. Its derivative P' has at least $m - 1$ positive roots.*

Proof. This is an easy, but not immediate, application of Rolle's Theorem. Suppose P has n distinct roots, s of them simple roots, and k of them multiple roots of orders $m_1, \ldots, m_k$, respectively. Then

$$m = s + m_1 + \ldots + m_k \quad \text{and} \quad n = s + k.$$

By Rolle's Theorem, $P'(X)$ has a root between each successive pair of distinct roots of P. Moreover, if P has an m_i-fold root at r_i, then P' has an $(m_i - 1)$-fold root there. Thus P' has at least

$$n - 1 + m_1 - 1 + \ldots + m_k - 1 = n - 1 + m_1 + \ldots + m_k - k$$
$$= n - k + m_1 + \ldots + m_k - 1$$
$$= s + m_1 + \ldots + m_k - 1 = m - 1$$

positive roots.   □

Proof of Theorem 2.4.i. By Lemma 2.7, we already know that the difference between the number of positive roots and the number of sign changes is even. We show the former number to be bounded by the latter by induction on the degree of P.

Basis. If $n = 0, P(X) = a_0 \neq 0$ has no positive roots and no sign changes. Thus there are no more roots than sign changes.

If $n = 1, P(X) = a_0 X + a_1$ either has no sign change and no positive root, or one sign change from a_0 to a_1 and one positive root, $-a_1/a_0$. Either way, there are no more roots than there are sign changes.

Induction Step. If the number m of positive roots of P exceeds the number of sign changes, the excess is at least 2. Applying Lemma 2.8, P' has at least

$$m - 1 > m - 2$$

roots, i.e. P' has at least one more root than P has sign changes. But P' has no more sign changes than P does. (Up to positive multiples, its coefficients are a truncation of the sequence of coefficients of P.) But the degree of P' is less than that of P and the induction hypothesis applies: the number of roots of P' is at most the number of its sign changes, which is at most $m - 2$, a contradiction.   □

As Theorem 2.4.ii follows immediately from Theorem 2.4.i, we have proven Theorem 2.4. Theorem 2.3, of course, follows as its two statements merely offer weaker conclusions than those of Theorem 2.4. Theorem 2.2.i is the same as 2.3.i and Theorem 2.5.i is the same as 2.4.i, which leaves us with the tasks of deducing 2.2.ii from 2.3.ii and 2.5.ii from 2.4.ii. We shall take a different tack with these latter by giving a different proof of these theorems.

Why an alternate proof? Any proof of Theorem 2.4 is going to require some use of the Calculus, specifically, an application of the Intermediate Value Theorem. For, unlike Theorems 2.2 and 2.3, Theorem 2.4 is an existence theorem:

2.9 Examples. i. Let $P(X) = X^3 - 1$. The sequence of coefficients of P is $1, 0, 0, -1$ and exhibits one sign change. Let m be the number of positive roots of P. Now $m \geq 0$, but Theorem 2.4 says $m \leq 1$ and $1 - m$ is even. This can only happen if $m = 1$: there is a positive root.
ii. Let $P(X) = X^3 - 2$. For the same reason as in part i, P has an unique positive root. This time, however, the root is not rational. In fact, it is not a constructible real number. Theorem 2.4 is not valid in the field of rational numbers, nor in that of constructible reals.
iii. Theorem 2.4 is valid for any real closed field. Now the real closed fields can be characterised as those ordered fields in which the Intermediate Value Theorem holds. It follows that the proof of Theorem 2.4 must depend on the Intermediate Value Theorem or some consequence thereof.

Theorems 2.2 and 2.3 do not have any existential import and hold for the rational numbers or, indeed, any subfield of the field of real numbers. They can be given elementary, Calculus-free proofs. Indeed, such proofs can be appealed to to yield Theorems 2.4 and 2.5 with a minimal reliance on the Intermediate Value Theorem. It is to such a proof that we next turn our attention.

The Calculus-based proof of theorem 2.4.i given defies convention in that the induction was on the degree of the polynomial P and went from P' to P. Following Segner, de Gua, Fourier, and Gauss, most proofs do an induction passing from a polynomial $P(X)$ to a polynomial $(X - r)P(X)$. If

$$P(X) = a_0 X^n + a_1 X^{n-1} + \ldots + a_{n-1} X + a_n,$$

then, for any real number r,

$$(X - r)P(X) = a_0 X^{n+1} + (a_1 - ra_0)X^n + \ldots + (a_n - ra_{n-1})X - ra_n.$$

Abstracting away the polynomials, the main thing one needs to prove is the following:

2.10 Lemma. *Assume $a_0 \neq 0$. Let c be the number of sign changes and p the number of permanences of the sequence*

$$a_0, \ldots, a_n. \tag{22}$$

Similarly, let c' be the number of sign changes and p' the number of permanences of the sequence

$$a_0, a_1 - ra_0, a_2 - ra_1, \ldots, a_n - ra_{n-1}, -ra_n, \tag{23}$$

where r is any real number.
i. if $r > 0$, $c' - c$ is a positive odd integer;
ii. if $r < 0$, $p' - p$ is a positive odd integer; and
iii. if $r = 0$, $p' = p + 1$ and $c' = c$.

I list the conclusions in diminishing order of importance. Theorems 2.2.i, 2.3.i, 2.4.i, and 2.5.i all require Lemma 2.10.i, while only Theorems 2.2.ii and 2.5.ii require Lemma 2.10.ii, and nothing depends on Lemma 2.10.iii.

Proof of Lemma 2.10. (I follow Krishnaiah's proof mentioned earlier.) Part iii is trivial as (23) simplifies to

$$a_0, a_1, \ldots, a_{n-1}, a_n, 0$$

when $r = 0$. All the transitions from a_i to a_{i+1} retain their character, and the new one from a_n to 0 is a permanence because, by definition, the final 0 inherits its sign from a_n.

i. Let $r > 0$. The proof is by induction on c.

Basis. $c = 0$, i.e. there are no sign changes in (22). Then $a_0, \ldots, a_n$ all have the same sign. Let k be the largest subscript of a nonzero term a_k in the sequence. Then (23) looks like

$$a_0, a_1 - ra_0, \ldots, a_k - ra_{k-1}, -ra_k, (0, \ldots, 0), \tag{24}$$

where $(0, \ldots, 0)$ denotes a possibly empty block of 0s. Because a_0 and $-ra_k$ have opposite signs, (24) has an odd number of sign changes.

Induction Step. Suppose (22) has $c > 0$ sign changes. Choose m so that the transition from a_{m-1} to a_m is the last sign change in the sequence. There are no changes in the tail $a_m, \ldots, a_n$.

Case 1. a_{m-1} is 0. Let k be the largest subscript $< m$ for which $a_k \neq 0$. Then (22) looks like

$$a_0, \ldots, a_k, 0, (0, \ldots, 0), a_m, \ldots, a_n,$$

where $(0, \ldots, 0)$ is a possibly empty block of 0s, and (23) looks like

$$a_0, a_1 - ra_0, \ldots, a_k - ra_{k-1}, -ra_k, (0, \ldots, 0), a_m, a_{m+1} - ra_m, \ldots, -ra_n. \tag{25}$$

Applying the induction hypothesis to $a_0, \ldots, a_k$, we see that $a_0, a_1 - ra_0, \ldots, -ra_k$ has $c - 1 + 2s + 1$ sign changes, where s is a nonnegative integer, and

$a_m, a_{m+1} - ra_m, \ldots, -ra_n$ has $2t + 1$ sign changes, where t is a nonnegative integer. Because a_k and a_m are of opposite sign, there are no changes of sign in the block $-ra_k, (0, \ldots, 0), a_m$. Hence (25) has a total of

$$c - 1 + 2s + 1 + 2t + 1 = c + 2(s + t) + 1$$

sign changes and $s + t \geq 0$.

Case 2. $a_{m-1} \neq 0$. Because a_{m-1} and a_m have opposite signs, $a_m - ra_{m-1}$ has the same sign as a_m and as $-ra_{m-1}$. Thus,

$$a_0, \ldots, a_{m-1} - ra_{m-2}, a_m - ra_{m-1}$$

has the same number of sign changes as

$$a_0, \ldots, a_{m-1} - ra_{m-2}, -ra_{m-1}$$

which is $c - 1 + 2s + 1$ for some nonnegative integer s. And

$$a_m - ra_{m-1}, a_{m+1} - ra_m, \ldots, -ra_n$$

has the same number of sign changes as

$$a_m, a_{m+1} - ra_m, \ldots, -ra_n,$$

which is $2t + 1$ for some nonnegative integer t. Thus the full sequence

$$a_0, a_1 - ra_0, \ldots, a_n - ra_{n-1}, -ra_n$$

has $c - 1 + 2s + 1 + 2t + 1 = c + 2(s + t) + 1$ sign changes with $2(s + t) + 1 > 0$.

ii. Let $r < 0$. The proof is by induction on p.

Basis. $p = 0$. Then $n = 0$ or the elements of (22) alternate in sign.

If $n = 0$, (23) looks like $a_0, -ra_0$, but since r is negative a_0 and $-ra_0$ have the same sign and we see $p' = 1$ and $p' - p = 1 > 0$.

If $n > 0$, again a_i and $-ra_i$ always have the same sign and this is the opposite of the sign of a_{i+1}. Hence $a_{i+1} - ra_i$ has the same sign as a_{i+1}. Hence sequence (23) alternates in sign at all but the last transition, where $a_n - ra_{n-1}$ has the same sign as $-ra_n$, which is the same as that of a_n. Again $p' = 1$.

Induction Step. Let $p > 0$ and assume the Lemma true for $p - 1$. Let the last permanence occur in the transition from a_{m-1} to a_m. The tail $a_m, \ldots, a_n$ either consists only of a_m or of nothing but sign changes.

Case 1. $a_m = 0$. Sequence (22) looks like

$$a_0, \ldots, a_{m-1}, 0, a_{m+1}, \ldots, a_n$$

with $a_{m+1} \neq 0$. (Why?) Sequence (23) looks like

$$a_0, a_1 - ra_0, \ldots, a_{m-1} - ra_{m-2}, -ra_{m-1}, a_{m+1}, a_{m+2} - a_{m+1}, \ldots, -ra_n. \quad (26)$$

By induction hypothesis there are $p - 1 + 2s + 1$ permanences in $a_0, a_1 - ra_0, \ldots, -ra_{m-1}$ and $2t + 1$ permanences in $a_{m+1}, \ldots, -ra_n$ for some nonnegative integers s, t. Because the transition from 0 to a_{m+1} was a sign change, a_{m-1} and a_{m+1} have opposite signs, whence so do $-ra_{m-1}$ and a_{m+1} and the total number of permanences in (26) is $p' = p - 1 + 2s + 1 + 2t + 1$, whence $p' - p > 0$ is odd.

Case 2. $a_m \neq 0$. Then sequence (23) looks like

$$a_0, a_1 - ra_0, \ldots, a_{m-1} - ra_{m-2}, a_m - ra_{m-1}, a_{m+1} - ra_m, \ldots, -ra_n.$$

Once again, $a_0 \ldots, a_{m-1}$ has $p - 1$ permanences, whence

$$a_0, a_1 - ra_0, \ldots, a_{m-1} - ra_{m-2}, -ra_{m-1} \tag{27}$$

has $p - 1 + 2s + 1$ permanences for some nonnegative integer s. But because the transition from a_{m-1} to a_m in (23) was a permanence, $a_{m-1}, -ra_{m-1}$, and $a_m - ra_{m-1}$ all have the same sign and

$$a_0, a_1 - ra_0, \ldots, a_{m-1} - ra_{m-2}, a_m - ra_{m-1}$$

has the same number, $p + 2s$, of permanences as (27). For the same reason,

$$a_m - ra_{m-1}, a_{m+1} - ra_m, \ldots, -ra_n$$

has the same number of permanences as

$$a_m, a_{m+1} - ra_m, \ldots, -ra_n,$$

which by induction hypothesis is $2t + 1$ for some nonnegative integer t. Thus, once again, $p' - p = 2(s + t) + 1 > 0$ is odd. $\qquad\qquad\square$

2.11 Remark. To complete the picture, I note that, under the assumptions of Lemma 2.10,
i. if $r > 0, p' - p \leq 0$ is even;
ii. if $r < 0, c' - c \leq 0$ is even.
For, there are n transitions in (22) and $n + 1$ in (23), all of which are either sign changes or permanences. Thus,

$$p' + c' = n + 1 = p + c + 1,$$

whence

$$p' - p = 1 - (c' - c). \tag{28}$$

2.12 Remark. The proof of the Lemma 2.10 simplifies ever so slightly if one only wishes to verify the differences $c' - c$ and $p' - p$ to be positive in their respective cases. If one does this, one can recover the missing information by noting that the sequences (22) and (23) have the same starting value a_0, but their end values are a_n and $-ra_n$, respectively. If $r > 0$, this means $c' - c$ is odd, and if $r < 0$, it means $c' - c$ is even, whence (28) tells us $p' - p$ is odd. For this argument, some authors like to replace the sequences $a_0, a_1, \ldots, a_n$ and $a_0, a_1 - ra_0, \ldots, a_n - ra_{m-1}, -ra_n$ by their respective sign sequences $s_0, \ldots, s_n$ and $t_0, \ldots, t_{n+1}$ and note that

$$(-1)^c = \prod_{i=0}^{n-1} \frac{s_i}{s_{i+1}} = \frac{s_0}{s_1} \cdot \frac{s_1}{s_2} \cdots \frac{s_{n-1}}{s_n} = \frac{s_0}{s_n}$$

$$(-1)^{c'} = \prod_{i=0}^{n} \frac{t_i}{t_{i+1}} = \frac{t_0}{t_1} \cdot \frac{t_1}{t_2} \cdots \frac{t_n}{t_{n+1}} = \frac{t_0}{t_{n+1}} = \frac{s_0}{t_{n+1}}$$

and that, as $t_{n+1} = -s_n$ if $r > 0$ and $t_{n+1} = s_n$ if $r < 0$,

$$(-1)^{c'} = \begin{cases} -(-1)^c, & r > 0 \\[2mm] (-1)^c, & r < 0. \end{cases}$$

2.13 Remark. Speaking of alternate proofs, one can prove Lemma 2.10 by induction on n, assuming the result for the sequence $a_0, \ldots, a_n$ and deriving it for $a_0, \ldots, a_n, a_{n+1}$. The basis step is easier: when $n = 0$, the sequence a_0 has no sign changes or permanences, while $a_0, -ra_0$ has one change if $r > 0$ and one permanence if $r < 0$. The induction step is handled similarly, but there are more cases to consider: $a_{n+1} = 0$; $a_{n+1} \neq 0$, but $a_n = 0$; neither a_{n+1} nor a_n is 0; each of the last two cases themselves subdividing according to whether the last transition is a sign change or a permanence.

2.14 Remark. The only property of r that was used in the proof was whether it was positive or negative, and this was only used locally. Hence, if for (23) we substituted the sequence

$$a_0, a_1 - r_0 a_0, \ldots, a_n - r_{n-1} a_{n-1}, -r_n a_n,$$

where $r_0, \ldots, r_n$ are all positive or all negative, Lemma 2.10 still holds: if c', p' now denote the numbers of sign changes and permanences, respectively, of this new sequence, then $c' - c$ is a positive odd integer when the r_i's are all positive and $p' - p$ is a positive odd integer when the r_i's are all negative.

But enough about the Lemma itself! As interesting an arithmetico-combinatorial result it may be, we proved it for the purpose of giving an elementary proof of Descartes' Rule of Signs.

Proof of Theorem 2.2. The proofs of parts i and ii are identical and proceed by induction on the number m of positive (respectively, negative) roots of P. We prove the result for the number of positive roots.

Basis. If P has no positive roots, then the number of such roots certainly does not exceed the number of sign changes.

Induction Step. If Q has $m + 1$ positive roots and r is one of them, then we can write

$$Q(X) = (X - r)P(X),$$

where $P(X) = a_0 X^n + \ldots + a_n$ has m positive roots. Assume P has c sign changes. Now, the sequence of coefficients of the polynomials P and Q are (22) and (23), respectively, whence Lemma 2.10 tells us that Q has at least $c + 1$ sign changes. But

$$m \leq c, \quad \text{by induction hypothesis,}$$

whence $m + 1 \leq c + 1$.   □

The proof of Theorem 2.5 is similar. The induction step requires a tiny calculation, but is otherwise handled in the same manner. The big difference between the proofs is the basis step, which is no longer a triviality, but an application of the Intermediate Value Theorem, which we may isolate as a pair of lemmas. For the positive case, there is the following.

2.15 Lemma. *Let* $P(X) = a_0 X^n + \ldots + a_n$ *be a polynomial possessing no positive real root. The number c of sign changes of P is even.*

Proof. Without loss of generality, we may assume $P(0) \neq 0$, i.e. $a_n \neq 0$. Multiplying $P(X)$ by -1 if necessary, we may also assume $a_0 > 0$. Then $P(x) > 0$ for all sufficiently large x. If $P(0) < 0$, then by the Intermediate Value Theorem P has a positive root, contrary to assumption. Hence $P(0) = a_n > 0$. But then the number of sign changes in the sequence $a_0, \ldots, a_n$ connecting the positive values a_0 and a_n has to be even.   □

Proof of Theorem 2.5.i. By induction on the number m of positive roots of the polynomial P.

Basis. By Lemma 2.15.

Induction Step. Let Q have $m + 1$ positive roots and write

$$Q(X) = (X - r)P(X),$$

where $r > 0, P$ has m positive roots and c sign changes as in the proof of Theorem 2.2. By induction hypothesis,

$$c = m + 2k, \quad \text{for some integer } k \geq 0.$$

By Lemma 2.10, if c' is the number of sign changes of Q, we have

$$c' = c + 2s + 1, \quad \text{for some integer } s \geq 0$$
$$= m + 2k + 2s + 1$$
$$= (m + 1) + 2(k + s),$$

whence $m + 1$ falls short of c' by an even integer.   □

The proof of Theorem 2.5.ii requires a more carefully stated lemma to serve as a basis for the corresponding induction.

2.16 Lemma. *Let P be a polynomial with no negative roots and suppose 0 is a root of P with multiplicity $z \geq 0$ (where $z = 0$ means 0 is not a root). Let p be the number of permanences of sign of P. Then: $p - z$ is even.*

Proof. By induction on the total number of roots P has.

Basis. P has no roots at all. By the Intermediate Value Theorem, the degree of P must be even. By Theorem 2.5.i, the number c of sign changes of P is also even. Thus, the number $p = n - c$ of permanences is the difference of two even numbers, hence itself even.

Induction Step. Suppose Q has $m + 1$ roots. We can write

$$Q = (X - r)P(X),$$

where $r \geq 0$. Let p, p' denote the numbers of permanences of P, Q, respectively, and z, z' the respective multiplicities of 0 as their roots.

If $r = 0, p' = p + 1$ and $z' = z + 1$, whence by induction hypothesis,

$$p' - z' = (p + 1) - (z + 1) = p - z$$

is even.

If $r > 0$, then by Remark 2.11, $p' - p$ is even. By the induction hypothesis, $p - z$ is even. Moreover, $z' = z$, whence

$$p' - z' = (p' - p) + (p - z') = (p' - p) + (p - z)$$

is the sum of two even numbers, hence is even.   □

Alternate Proof. By the Fundamental Theorem of Algebra,

$$P = \prod_{i=1}^{k} Q_i \cdot \prod_{i=1}^{m} L_i \cdot X^z,$$

where each Q_i is a quadratic with no real root, and each L_i is linear with a positive root. Then

$$deg(P) = 2k + m + z.$$

For c the number of sign changes in P, we know $c - m$ is even by Theorem 2.5.i. Now

$$p + c = deg(P) = 2k + m + z,$$

whence

$$p - z = 2k + m - c = 2k - (c - m)$$

is even.   □

The inductive proof of Theorem 2.5.ii may now be given. As the proof is completely analogous to that of 2.5.i, I omit the details.

From Theorem 2.5 the other versions of Descartes' Rule of Signs follow as corollaries. This does not, however, make Theorem 2.5 the best result.

2.17 Example. Let $P(X) = X^6 - X^3 + 1$. The sequence of coefficients is $1, 0, 0, -1, 0, 0, 1$. There are 4 permanences of sign and 0 is not a root, whence the number of negative roots according to Theorem 2.5.ii is 0, 2, or 4. However, $Q(X) = P(-X)$ has the sequence $1, 0, 0, 1, 0, 0, 1$ with no sign changes, whence Theorem 2.4.ii tells us P has no negative roots at all.

In this Example, 2.4.ii yielded a sharper estimate than 2.5.ii. This is not always the case, as the example $P(X) = X^2 - X + 1$ shows. In this case, both results tell us there are no negative roots. I cannot give an example, however, in which 2.4.ii gives a worse estimate than 2.5.ii: the upper bound given by 2.4.ii is never greater than that given by 2.5.ii. To prove this, we have to take a closer look at the passage from the polynomial $P(X)$ to $Q(X) = P(-X)$.

The following special case is more-or-less obvious.

2.18 Lemma. *Let $P(X)$ have only nonzero coefficients and define $Q(X) = P(-X)$. Let c, c' denote the respective number of sign changes of P and Q, and p, p' their respective numbers of permanences. Then*
i. $c' = p$;
ii. $p' = c$.

Proof. Let (22) be the sequence of coefficients of P. Then the sequence of coefficients of Q is

$$a_0, -a_1, a_2, -a_3, \ldots, -a_{n-1}, a_n \quad \text{if } n \text{ is even}$$

$$-a_0, a_1, -a_2, a_3, \ldots, -a_{n-1}, a_n \quad \text{if } n \text{ is odd.}$$

Clearly every sign change from a_i to a_{i+1} is a permanence from $-a_i$ to a_{i+1} or from a_i to $-a_{i+1}$, whichever occurs in the sequence of coefficients of Q, and every permanence from a_i to a_{i+1} becomes a sign change. $\qquad\square$

If one of a_i and a_{i+1} is 0, the resulting transition in the sequence of coefficients of Q may not be of the opposite type. For the polynomial $X^6 - X^3 + 1$ of Example 2.17, which had 4 permanences and 2 sign changes, we did not finish up with 2 permanences and 4 sign changes because of the 0s.

Examining the situation, we see that there are four distinct types of transitions between successive nonzero entries a and b in a sequence (22):

Type I. There is an even number (possibly zero) of 0s between a and b, and a, b have the same sign. We let

$$E^p = \text{the number of transitions of this type.} \tag{29}$$

Type II. There is an even number (possibly zero) of 0s between a and b, and a, b have opposite signs. We let

$$E^c = \text{the number of transitions of this type.} \tag{30}$$

Type III. There is an odd number of 0s between a and b, and a, b have the same sign. We let

$$O^p = \text{the number of transitions of this type.} \tag{31}$$

Type IV. There is an odd number of 0s between a and b, and a, b have opposite signs. We let

$$O^c = \text{the number of transitions of this type.} \tag{32}$$

Additionally, let

$$e^p = \text{the total number of 0s occurring in blocks of Type I} \tag{33}$$
$$e^c = \text{the total number of 0s occurring in blocks of Type II} \tag{34}$$
$$o^p = \text{the total number of 0s occurring in blocks of Type III} \tag{35}$$
$$o^c = \text{the total number of 0s occurring in blocks of Type IV.} \tag{36}$$

Finally, let

$$\nu = e^p + e^c + o^p + o^c \tag{37}$$

be the total number of 0s trapped between nonzero entries in the sequence of coefficients of P.

The following Lemma is essentially due to Gauss.

2.19 Lemma. *Let* $P(X) = a_0 X^n + \ldots + a_n$ *be given and let* $Q(X) = P(-X)$. *Let* c, c' *denote the number of sign changes in* P *and* Q, *respectively, and let* p, p' *denote their respective numbers of permanences. Finally, let* z *be the common multiplicity of* 0 *as a root of both polynomials. Then:*

$$p = E^p + O^p + \nu + z \qquad\qquad p' + E^c + O^p + \nu + z$$
$$c = E^c + O^c \qquad\qquad\qquad c' = E^p + O^c.$$

Proof. First note that $\nu + z = e^p + e^c + o^p + o^c + z$ counts the total number 0s in the sequences of coefficients of the two polynomials. They count the immediate transitions from numbers to 0, and hence count permanences of this kind. Thus $\nu + z$ is a summand of both p and p'.

We can think of E^p as counting transitions from 0 to b at the end of even blocks of 0s (or, from a to b if the block has no 0s) that are permanences in $a_0, \ldots, a_n$, and O^p as counting such transitions at the end of odd blocks of 0s. These numbers added to $\nu + z$ yield p as we've exhausted all the permanences of P.

Similarly, $E^c + O^c$ counts all the sign changes of P.

As for p' and c', consider the effect the transformation from $P(X)$ to $Q(X)$ has on one of these blocks. For convenience, let us write 0^k to denote a block of k 0s.

Type I. $a0^{2r}b$, with a, b sharing a sign, transforms into $\pm a0^{2r} \mp b$. The last transition has become a sign change. The overall contribution is E^p to c'.

Type II. $a0^{2r}b$, with a, b of opposite signs, again transforms into $\pm a0^{2r} \mp b$. The last transition has become a permanence. The overall contribution is E^c

to p'.

Type III. $a0^{2r+1}b$, with a, b sharing a sign, transforms into $\pm a0^{2r+1} \pm b$. The last transition remains a permanence. The overall contribution is O^p to p'.

Type III. $a0^{2r+1}b$, with a, b of opposite signs, transforms into $\pm a0^{2r+1} \pm b$. But $\pm a$ and $\pm b$ still have opposite signs. The overall contribution is O^c to c'. $\square$

2.20 Corollary. *Let $Q(X) = P(-X)$ and suppose P has p permanences and Q has c' sign changes. Then: $c' \leq p - z$. In particular, the bound c' on the number of negative roots of P given by Theorem 2.4.ii is never worse than the bound $p - z$ given by Theorem 2.5.ii.*

Proof. Using the notation of Lemma 2.19,

$$
\begin{aligned}
p - z &= E^p + O^p + e^p + e^c + o^p + o^c \\
&\geq E^p + o^c \\
&\geq E^p + O^c = c'.
\end{aligned}
$$

$\square$

Gauss originally applied Lemma 2.19 to prove the following:

2.21 Corollary. *Let $P(X)$ be as in the statement of Lemma 2.19. The number of complex roots of P is at least $O^p - O^c + \nu$.*

Proof. Let m^+, m^- denote the number of positive and negative roots, respectively, of P. Further let z denote the multiplicity of 0 as a root of P. Then $m^+ + m^- + z$ is the total number of real roots, whence the number of complex roots of P is

$$
\begin{aligned}
n - m^+ - m^- - z &= p + c - m^+ - m^- - z \\
&\geq p + c - c - c' - z, \text{ by Theorem 2.2} \\
&\geq p - c' - z \\
&\geq O^p - O^c + \nu \text{ by Lemma 2.19.}
\end{aligned}
$$

$\square$

2.22 Corollary. *Let $P(X) = a_0 X^n + \ldots + a_n$. If there is a k such that*

$$
a_k > 0, a_{k+1} = 0, a_{k+2} > 0 \quad \text{or} \quad a_k < 0, a_{k+1} = 0, a_{k+2} < 0,
$$

then $P(X)$ has a pair of complex roots.

Proof. By Corollary 2.21, the number of complex roots is at least

$$
\begin{aligned}
O^p - O^c + \nu = O^p - O^c + e^p + e^c + o^p + o^c \\
\geq O^p - O^c + o^p + o^c \\
\geq O^p + o^p, \text{ since } o^c \geq O^c \\
\geq 2,
\end{aligned}
$$

if $a_k > 0, a_{k+1} = 0, a_{k+2} > 0$ or $a_k < 0, a_{k+1} = 0, a_{k+2} < 0$. $\square$

Corollary 2.22 is de Gua's result referred to by Horner, and this proof certainly is elementary, but it hardly yields any insight into why the result should be true. For that we turn to the Calculus.

We will take a closer look at de Gua's Theorem in the next section. For now, let us take another look at Euler's, Lagrange's, and Newton's equations and see what Descartes' Rule alone can tell us.

2.23 Example. i. Euler's polynomial,

$$P(X) = X^4 - 4X^3 + 8X^2 - 16X + 20 = 0,$$

has 4 sign changes and no permanences. All roots are positive and there are 0, 2, or 4 of them.

$$P_1(X) = P(X + 1) = X^4 + 2X^2 - 8X + 9$$

has 2 sign changes, while

$$P_1(-X) = X^4 + 2X^2 + 8X + 9$$

has no sign changes. All roots of P_1 are positive and there are 0 or 2 of them. Thus P has 0 or 2 roots greater than 1. Finally,

$$P_2(X) = P(X + 2) = X^4 + 4X^3 + 8X^2 + 4$$

has no sign changes, while

$$P_2(-X) = X^4 - 4X^3 + 8X^2 + 4$$

has 2 sign changes. This means P has 0 or 2 roots less than 2. Summarising, P either has no roots or it has 2 roots between 1 and 2.

ii. We haven't completely exhausted the usefulness of Descartes' Rule. For we can consider the general

$$P_{1+r}(X) = P_1(X + r) = X^4 + 4rX^3 + (2 + 6r^2)X^2 \\ + (-8 + 4r + 4r^3)X + (9 - 8r + 2r^2 + r^4). \tag{38}$$

For $0 < r < 1$, the first three coefficients of (38) are positive, the fourth negative, and the last of undetermined sign. Now, if $1 + r$ is a root of P, i.e.

$$0 = P(1 + r) = P_{1+r}(0) = P_1(r),$$

the last coefficient of (38) is 0 and $P_{1+r}(X)$ has exactly one sign change. Thus $P_{1+r}(X)$ has a positive root, i.e. P has a root greater than $1 + r$. This cannot happen if $1 + r$ is the greater of the 2 possible roots of P in the interval $[1, 2]$. Hence P has no real roots.

iii. Another way of seeing P has no real roots is to rewrite, say, P_2 as

$$P_2(X) = X^4 + 4X^3 + 4X^2 + 4X^2 + 4$$
$$= (X^2 + 4X + 4)X^2 + 4X^2 + 4$$
$$= (X + 2)^2 X^2 + 4X^2 + 4,$$

a sum of squares, one of which is never 0.

iv. Yet another approach is to tackle P directly:

$$P(X) = X^4 - 4X^3 + 4.5X^2 + 3.5X^2 - 16X + 20$$
$$= (X^2 - 4X + 4.5)X^2 + (3.5X^2 - 16X + 20).$$

Each of $X^2 - 4X + 4.5$ and $3.5X^2 - 16X + 20$ has a negative discriminant:

$$(-4)^2 - 4(1)(4.5) = -2 < 0$$
$$(-16)^2 - 4(3.5)(20) = -24 < 0.$$

Hence both quadratics are always positive and so is P.

2.24 Example. i. Lagrange's equation,

$$P(X) = X^3 - 7X + 7 = 0,$$

has 2 sign changes and 1 permanence, whence P has 0 or 2 positive roots and 1 negative root. The problem was to locate the positive roots. Once again we look at

$$P_1(X) = P(X + 1) = X^3 + 3X^2 - 4X + 1,$$

which has 2 sign changes and 1 permanence. This means P has 0 or 2 roots greater than 1 and 1 root less than 1, which we already know to be negative. Finally,

$$P_2(X) = P(X + 2) = X^3 + 6X^2 + 5X + 1$$

has no sign changes and

$$P_2(-X) = -X^3 + 6X^2 - 5X + 1$$

has 3 sign changes. Thus P has no roots greater than 2 and 1 or 3 roots less than 2. This doesn't tell us P has any positive roots, but it does narrow the search to the interval $[1, 2]$.

ii. We can repeat the trick of part ii of the previous Example by considering $P_{1+r}(X)$ for $0 < r < 1$:

$$P_{1+r}(X) = X^3 + (3r + 3)X^2$$
$$+ (3r^2 + 6r - 4)X + (r^3 + 3r^2 - 4r + 1).$$

The first two coefficients are positive for $0 < r < 1$, the third changes sign from negative to positive around .527. So we try $r = .5$, say, and observe

$$P_{1.5}(X) = X^3 + 4.5X^2 - .25X - .125.$$

This has 1 sign change, whence 1 positive root and, without looking at $P_{1.5}(-X)$, two negative roots. Thus P has one negative root, one root in the interval $[1, 1.5]$, and one in the interval $[1.5, 2]$.

2.25 Example. i. Newton's equation,

$$P(X) = X^3 - 2X - 5 = 0,$$

has one sign change, and

$$P(-X) = -X^3 + 2X - 5 = 0$$

has two sign changes. Hence P has 1 positive and 0 or 2 negative roots. The positive root we've accounted for. Does the equation have any negative roots?

$$P_{-1}(X) = P(X - 1) = X^3 - 3X^2 + X - 4$$

has 3 sign changes and no permanences, whence any negative root of P is greater than -1.

ii. We can look more generally at

$$P_r(X) = P(X + r) = X^3 + (3r)X^2 + (3r^2 - 2)X + (r^3 - 2r - 5),$$

for $-1 < r < 0$. The first coefficient is positive, the second negative, and the remaining two unknown. However, $3r^2 - 2 = 0$ for $r = -\sqrt{2/3} \approx -.816$ and there we have

$$P_{-.816...}(X) = X^3 - 2.449X^2 - 3.911 \tag{39}$$
$$P_{-.816...}(-X) = -X^3 - 2.449X^2 - 3.911,$$

and we see P has one root greater than $-.816\ldots$, and none smaller.

Note that, from the expression (39) for $P_{-.816...}(X)$, we could as easily have appealed to de Gua's Theorem (i.e. Corollary 2.22) to conclude Newton's polynomial to have no roots other than that near 2. If we were not aware of de Gua's Theorem, however, the necessary extra step of looking at $P_{-.816...}(-X)$ is a small one. The real advantage is not afforded by de Gua's Theorem as we've proven it, but by Horner's more general statement of it, as in the example of Euler's equation.

3 De Gua's Theorem

The appellation "De Gua's Theorem" is a misnomer. Horner credited the observation I have been calling "De Gua's Theorem" to de Gua and, lacking a better name for it, I have simply chosen the eponymous designation. The result, as can be gleaned from Bartolozzi and Franci's history of Descartes' Rule cited in the preceding section, is older, going back at least to Segner's first partial proof of the Theorem in 1728 and given a full proof by George Campbell in 1729.

Campbell's proof is rather interesting. He starts with a polynomial

$$P(X) = \ldots + a_k X^{n-k} + a_{k+1} X^{n-k-1} + a_{k+2} X^{n-k-2} + \ldots,$$

all of the roots of which are real. If all the roots of a polynomial are real, then all the roots of its derivative are real (following Lemma 2.8 of the preceding section). Differentiate P $n - k - 2$ times to get

$$Q(X) = \ldots + (n-k)\cdots 3a_k X^2 + (n-k-1)\cdots 2a_{k+1}X + (n-k-2)!a_{k+2}$$

$$= \ldots + \frac{(n-k)!}{2!}a_k X^2 + \frac{(n-k-1)!}{1!}a_{k+1}X + (n-k-2)!a_{k+2}.$$

If we now reverse the order of the coefficients, we get

$$R(X) = X^{k+2}Q(\frac{1}{X}) = (n-k-2)!a_{k+2}X^{k+2}$$

$$+ \frac{(n-k-1)!}{1!}a_{k+1}X^{k+1} + \frac{(n-k)!}{2!}a_k X^k + \ldots,$$

which also has all real roots. (Why?) If we now differentiate this k times, we get

$$S(X) = \frac{(n-k-2)!\,(k+2)!}{0!\,2!}a_{k+2}X^2 + \frac{(n-k-1)!\,(k+1)!}{1!\,1!}a_{k+1}X$$

$$+ \frac{(n-k)!\,k!}{2!\,0!}a_k,$$

with all real roots. But $S(X)$ is quadratic, whence its discriminant is nonnegative:

$$((n-k-1)!(k+1)!a_{k+1})^2 \geq 4\frac{(n-k-2)!\,(k+2)!}{1}\frac{(n-k)!\,k!}{2}\frac{}{1}a_k a_{k+2},$$

whence

$$(a_{k+1})^2 \geq \frac{(k+2)(n-k)}{(k+1)(n-k-1)}a_k a_{k+2}.$$

In particular, if $a_{k+1} = 0$, a_k and a_{k+2} cannot have the same sign:

3.1 Theorem. *Let $P(X) = a_0 X^n + \ldots + a_n$ have n real roots. For no k does one have*

$$a_k > 0, a_{k+1} = 0, a_{k+2} > 0 \quad or \quad a_k < 0, a_{k+1} = 0, a_{k+2} < 0.$$

If we make a deeper use of the Calculus, we can give the Theorem a proof of less serendipitous appearance.

Alternate Proof. By Taylor's Theorem,

$$a_k = \frac{P^{(n-k)}(0)}{(n-k)!}, \quad a_{k+1} = \frac{P^{(n-k-1)}(0)}{(n-k-1)!} \quad a_{k+2} = \frac{P^{(n-k-2)}(0)}{(n-k-2)!},$$

whence the condition, say,

$$a_k > 0, \quad a_{k+1} = 0, \quad a_{k+2} > 0,$$

translates in reverse order to

$$f(0) > 0, \quad f'(0) = 0, \quad f''(0) > 0,$$

where $f(x) = P^{(n-k-2)}(x)$. This means that f has a local minimum at 0. Now f itself is a polynomial of degree $k+2$, where n is the degree of P. To show P has fewer than n roots, it suffices to show f to have fewer than $k+2$ roots. For, by the proof of Lemma 2.8, if Q' is the derivative of a polynomial Q, and Q' has only m real roots, Q cannot have more than $m+1$ real roots. By induction, Q cannot have more than j real roots more than $Q^{(j)}$.

Suppose then that all $k+2$ roots of f are real.

Case 1. All roots of f are positive. Listed with multiplicity, the roots are $r_1 \leq r_2 \leq \ldots \leq r_{k+2}$. By the proof of Lemma 2.8, there are $k+1$ positive roots $s_1 \leq \ldots \leq s_{k+1}$ of f', each s_i sandwiched between r_i and r_{i+1}. But $f'(0) = 0$ as well and the number of roots of f' exceeds $k+1$, a contradiction.

Case 2. All roots of f are negative. The analogous argument applies.

Case 3. $r_j < 0 < r_{j+1}$ for two successive roots r_j, r_{j+1} of f. Because these are successive roots and $f(0) > 0$, f is positive on $[r_j, r_{j+1}]$. (Here we use the Intermediate Value Theorem again.) Let f assume its maximum value on this interval at some point a. a differs from r_j, r_{j+1} because $f(a)$ is positive and f is 0 at the endpoints. Moreover, $a \neq 0$ because $f(0)$ is a local minimum. But $f'(a) = 0$ and we have at least two roots of f' in the interval $[r_j, r_{j+1}]$ and at least one between every other pair r_i and r_{i+1}. Adding them up yields $k+2$ roots of f', again one too many. □

The alternate proof brings out more clearly the local nature of the phenomenon than does Campbell's proof or the Gaussian proof given in the preceding section. For, it shows that the existence of such sequences a_k, a_{k+1}, a_{k+2} with, say,

$$a_k > 0, a_{k+1} = 0, a_{k+2} > 0$$

is associated with a missing pair of roots surrounding 0. If two distinct translated polynomials, $P_a(X) = P(X+a)$ and $P_b(X) = P(X+b)$ exhibit such sequences among their respective coefficients, then each of them is missing a pair of real roots on either side of 0— which translates back to missing pairs of roots of P sandwiching a and b, respectively. This makes plausible Horner's claim that these are two distinct pairs of roots, but it doesn't prove the claim.

Before giving a statement and proof of Horner's version of the de Gua-Campbell result, it might be instructive to consider a few examples. Some terminology may be in order.

3.2 Definitions. *Let* $P(X) = a_0 X^n + \ldots + a_n$ *be a given polynomial with* $a_0 \neq 0$. *Three successive coefficients* a_k, a_{k+1}, a_{k+2} *form a* de Gua triple *if*

$$a_k > 0, a_{k+1} = 0, a_{k+2} > 0 \quad or \quad a_k < 0, a_{k+1} = 0, a_{k+2} < 0.$$

A real number a *is a* de Gua point *of* P *if* $P_a(X) = P(X + a)$ *has a de Gua triple in its sequence of coefficients.*

If a and b are distinct de Gua points of P, we cannot say that they indicate distinct pairs of missing roots around a and b, respectively, for a and b could conceivably be trapped within a single pair of missing roots. Since the missing roots don't exist, there is no way of comparing the pairs. The complex roots, however, do exist and if we could associate the complex roots with the de Gua points, we might be able to show that distinct de Gua points belong to distinct complex roots. But I don't see how:

3.3 Example. i. Let $a + bi$ be any complex number which is not real, i.e. $b \neq 0$. We can construct a polynomial P having 0 as a de Gua point and having $a \pm bi$ among its roots. Start with

$$(X - (a + bi))(X - (a - bi)) = X^2 - 2aX + a^2 + b^2,$$

and multiply by a polynomial yet to be determined:

$$(X^2 - 2aX + a^2 + b^2)(d_0 X^m + \ldots + d_m). \tag{40}$$

The coefficient of

$$\begin{array}{lll} X^k & \text{is} & 1 \cdot d_{n-k-2} - 2a d_{n-k-1} + (a^2 + b^2) d_{n-k} & (41) \\ X^{k+1} & \text{is} & 1 \cdot d_{n-k-1} - 2a d_{n-k} + (a^2 + b^2) d_{n-k+1} & (42) \\ X^{k+2} & \text{is} & 1 \cdot d_{n-k} - 2a d_{n-k+1} + (a^2 + b^2) d_{n-k+2}. & (43) \end{array}$$

We can choose d_{n-k}, d_{n-k+1} arbitrarily and use (42) to determine d_{n-k-1}. After that, we have but to choose $d_{n-k\pm2}$ large enough to make (41), (43) positive. The resulting polynomial (40) has 0 as a de Gua point and $a \pm bi$ among its roots.

ii. Take $m = 4$ in (40), $k = 2$. Choose d_2, d_3 arbitrarily,

$$d_1 = 2a d_2 - (a^2 + b^2) d_3 \text{ by (42)}$$
$$d_0 > 2a d_1 - (a^2 + b^2) d_2 \text{ by (41)}$$
$$d_4 > \frac{2a d_3 - d_2}{a^2 + b^2} \text{ by (43)}.$$

For example, let $a = b = 1, d_2 = d_3 = 1$. Then $d_1 = 2 \cdot 1 \cdot 1 - (1^2 + 1^2) \cdot 1 = 0, d_0 > -2, d_4 > \frac{1}{2}$. If we choose $d_0 = d_4 = 1$, (40) becomes

$$P(X) = (X^2 - 2X + 2)(X^4 + X^2 + X + 1)$$
$$= X^6 - 2X^5 + 3X^4 - X^3 + X^2 + 2.$$

This has 0 as a de Gua point (note the missing linear term). It also has $a = \frac{1}{3}$ and $b \approx .0976831855$ as de Gua points. Which one is properly associated with the pair $1 \pm i$ of complex roots?

To find and create de Gua points other than 0, start with a polynomial,

$$P(X) = a_0 X^n + \ldots + a_n,$$

and a real number r, and consider the substitution

$$P_r(X) = P(X + r) = b_0(r)X^n + b_1(r)X^{n-1} + \ldots + b_n(r).$$

Each $b_i(r)$ is a polynomial in r. In fact, Taylor's Theorem even tells us

$$b_n(r) = P(r)$$
$$b_{n-1}(r) = P'(r) = b_n'(r)$$
$$b_{n-2}(r) = \frac{P''(r)}{2!} = \frac{b_{n-1}'(r)}{2}$$
$$b_{n-3}(r) = \frac{P^{(3)}(r)}{3!} = \frac{b_{n-2}'(r)}{3}$$

$$\vdots$$

$$b_{n-k+1}(r) = \frac{b_{n-k}'(r)}{k+1}$$

$$\vdots$$

$$b_0(r) = a_0.$$

3.4 Example. i. For the polynomial

$$P(X) = X^6 - 2X^5 + 3X^4 - X^3 + X^2 + 2$$

of Example 3.3.ii, we have

$$b_6(r) = r^6 - 2r^5 + 3r^4 - r^3 + r^2 + 2$$
$$b_5(r) = 6r^5 - 10r^4 + 12r^3 - 3r^2 + 2r$$
$$b_4(r) = 15r^4 - 20r^3 + 18r^2 - 3r + 1$$
$$b_3(r) = 20r^3 - 20r^2 + 12r - 1$$
$$b_2(r) = 15r^2 - 10r + 3$$
$$b_1(r) = 6r - 2$$
$$b_0(r) = 1.$$

Locating the de Gua points is a matter of finding the roots of $b_1, \ldots, b_5$, and evaluating $b_{i-1}(r)$ and $b_{i+1}(r)$ to check the signs when a root r of b_i is found. This is not as difficult as it sounds.

ii. $b_1(r)$ is always linear and has a root. In this case it is $r_1 = 1/3$, both $b_0(1/3)$ and $b_2(1/3)$ are positive, whence $r_1 = 1/3$ is a de Gua point.

iii. $b_2(r)$ is quadratic with negative discriminant and has no root.

iv. For $b_3(r)$ we can use our knowledge of the cubic equation. Substituting $t = r + \frac{1}{3}$ results in

$$b_3(r) = 20t^3 + \frac{16}{3}t + \frac{41}{27}.$$

We could now check the discriminant of this equation to verify that it has one real root, or we can appeal to the form of de Gua's Theorem we have already proven. Either way, b_3 has an unique real root, which we can approximate as .0976831855. Both b_2 and b_4 are positive at this argument, whence it is a de Gua Point.

v. Because b_3 is essentially the derivative of b_4 and b_4 is positive at .097..., where it assumes a minimum value, we conclude $b_4(r)$ to have no roots at all.

vi. This brings us to b_5, of which 0 is obviously a root and in fact a de Gua point. That b_5 has no further roots is perhaps most easily seen by writing

$$\frac{b_5(r)}{r} = 6r^4 - 10r^3 + 12r^2 - 3r + 2$$

$$= (6r^2 - 10r + 10)r^2 + (2r^2 - 3r + 2)$$

and noting the quadratics $6r^2 - 10r + 10$ and $2r^2 - 3r + 2$ to have negative discriminants.

3.5 Example. Following Example 3.3.i, choose $a = 0, b = 1$ so that $\pm i$ will be roots of our final equation. We may choose d_2, d_3 arbitrarily, and require

$$d_1 = -d_3, \quad d_0 > -d_2, \quad d_4 > -d_2.$$

Thus, let us choose $d_2 = 1, d_3 = -1, d_1 = 1, d_0 = 1, d_4 = 0$:

$$Q(X) = (X^2 + 1)(X^4 + X^3 + X^2 - X)$$
$$= X^6 + X^5 + 2X^4 + X^2 - X$$
$$= X(X^5 + X^4 + 2X^3 + X - 1).$$

So consider

$$P(X) = X^5 + X^4 + 2X^3 + X - 1.$$

Again, look at

$$P_r(X) = P(X + r) = X^5 + (5r + 1)X^4 + (10r^2 + 4r + 2)X^3$$
$$+ (10r^3 + 6r^2 + 6r)X^2$$
$$+ (5r^4 + 4r^3 + 6r^2 + 1)X$$
$$+ (r^5 + r^4 + 2r^3 + r - 1).$$

$b_1(r)$ has root $r = -1/5$,

$$P_{-1/5}(X) = X^5 + 1.6X^3 - 1.04X^3 + 1.216X - 1.215,$$

and $-1/5$ is a de Gua point. It is also clear that 0 is a de Gua point. The polynomials $b_1, \ldots, b_4$ have no other roots, whence there are no other de Gua points.

3.6 Example. We can also use the polynomials b_i to construct polynomials with some prescribed de Gua points. Suppose I want de Gua points at 0 and 2. I can choose them to be positive local minima of, say, $b_{k+2}(r)$. Between two such minima there must be a local maximum, say at 1. The simplest thing to do is to choose

$$b_{k+1}(r) = (r - 0)(r - 1)(r - 2) = r^3 - 3r^2 + 2r.$$

Then, up to a positive multiplicative constant,

$$b_{k+2}(r) = \frac{r^4}{4} - r^3 + r^2 + c_1$$

$$b_{k+3}(r) = \frac{r^5}{20} - \frac{r^4}{4} + \frac{r^3}{3} + c_1 r + c_2,$$

and, changing r to X and multiplying by 60,

$$P(X) = 3X^5 - 15X^4 + 20X^3 + C_1 X + C_2$$

for some constants C_1, C_2. If we calculate the true b_i's, we obtain successively

$$b_5(r) = 3r^5 - 15r^4 + 20r^3 + C_1 r + C_2$$

$$b_4(r) = 15r^4 - 60r^3 + 60r^2 + C_1 = 60(\frac{r^4}{4} - r^3 + r^2 + c_1)$$

$$b_3(r) = 30r^3 - 90r^2 + 60r = 30(r^3 - 3r^2 + 2r)$$

$$b_2(r) = 30r^2 - 60r + 20$$

$$b_1(r) = 15r - 15$$

$$b_0(r) = 3.$$

Then one readily checks that:

$$
\begin{array}{ll}
b_2(0) = 20 > 0 & b_2(2) = 30 \cdot 4 - 60 \cdot 2 + 20 = 20 > 0 \\
b_3(0) = 0 & b_3(2) = 30(8 - 3 \cdot 4 + 2 \cdot 2) = 0 \\
b_4(0) = C_1 & b_4(2) = 15 \cdot 16 - 60 \cdot 8 + 60 \cdot 4 + C_1 = C_1,
\end{array}
$$

and we see that $0, 2$ are de Gua points of $P(X)$ for any choice of $C_1 > 0$. C_2 may then be chosen arbitrarily. For example, if we choose $C_1 = 2$, then a choice of $C_2 = 40$ will give -1 as a root, while $C_2 = -10$ will make 1 the root.

Let us get down to proving Horner's assertion about de Gua points.

3.7 Definitions. *Let $P(X) = a_0 X^n + \ldots + a_n$ be a given polynomial, with $a_0 \neq 0$. A de Gua triple a_k, a_{k+1}, a_{k+2} is of order $n - (k+1)$, i.e. the order of the triple is the degree of the "missing" term $a_{k+1} X^{n-k-1}$. A de Gua point a is of order m if $P_a(X) = P(X + a)$ has a de Gua triple of order m.*

Note that i. the possible orders of de Gua triples are $1, 2, \ldots, n-1$; and ii. a de Gua point a will have several orders if P_a has several de Gua triples. We can define the *multiplicity* of a de Gua point a to be the total number of de Gua sequences the polynomial P_a has. When counting the number of de Gua points of a polynomial, as with counting the number of roots, multiplicity is assumed.

3.8 Theorem. *Let $P(X) = a_0 X^n + \ldots + a_n$ be a given polynomial of degree n. Let P have a total of d de Gua points. The number of real roots of P is at most $n - 2d$.*

Proof. By induction on n.

Basis. $n \leq 2$. If $n = 1$, P is linear and has an unique root. Moreover, P has no de Gua points because the polynomials P_a all have only 2 coefficients. Hence the upper bound $n - 2d = 1 - 0 = 1$ works.

If $n = 2$, $P(X) = a_0 X^2 + a_1 X + a_2$ and any de Gua point r has order 1. But

$$b_1(r) = 2a_0 r + a_1$$

has only one solution, whence there can be at most 1 de Gua point: $d \leq 1$.

If $d = 0$, $n - 2d = 2 - 0 = 2$ and a quadratic has at most 2 real roots.

If $d = 1$, $n - 2d = 2 - 2 = 0$. But, if r is the de Gua point, $P_r(X)$ is of the form $b_0 X^2 + b_2$, with b_0, b_2 both positive or both negative. In either case, P_r has no real root, whence neither does P.

Induction Step. $n > 2$. Notice that, for any r,

$$\left(\frac{dP}{dX}\right)_r (X) = \frac{dP_r}{dX}(X) = nb_0(r)X^{n-1} + \ldots + 2b_{n-2}(r)X + b_{n-1}(r),$$

whence

$$(n-k)b_k(r), (n-k-1)b_{k+1}(r), (n-k-2)b_{k+2}(r)$$

is a de Gua sequence of P_r' just in case $b_k(r), b_{k+1}(r), b_{k+2}(r)$ is a de Gua sequence of P_r with $k + 2 \leq n - 1$. In other words, the de Gua points of order m of P' for $1 \leq m \leq n - 2$ are precisely the de Gua points of order $m + 1$ of P. The new de Gua points in the passage from P' to P are those of order 1.

Letting

$$d_i = \text{ the number of de Gua points of order } i \text{ of } P,$$

and

$$e = \text{ the number of real roots of } P',$$

our induction hypothesis is

$$e \leq n - 1 - 2 \sum_{i=2}^{n-1} d_i,$$

whence it suffices to exstablish that

the number of real roots of $P \leq e + 1 - 2d_1$.

This is just an application of counting and Rolle's Theorem. Suppose P has m real roots and d_1 de Gua points of order 1. List all of these points in order from left to right, multiple roots being several times listed:

$$p_1 \leq p_2 \leq \ldots \leq p_k.$$

Claim. P' has a root between p_i and p_{i+1}, the inclusion being strict in case at least one of p_i and p_{i+1} is a de Gua point.

Proof. If p_i and p_{i+1} are roots, Lemma 2.8 applies.

If p_i is a root and p_{i+1} is a de Gua point, then p_{i+1} is a local extremum, either a positive minimum or a negative maximum. Suppose, for example, the former to be the case. Then somewhere strictly between p_i and p_{i+1} a local maximum occurs. P' is 0 at this point.

The same reasoning applies if p_i is a de Gua point and p_{i+1} a root.

If p_i and p_{i+1} are de Gua points, they are either both positive local minima or negative local maxima (else one is positive and one negative and they are separated by a root). If they are both minima, P has a local maximum between them; if both maxima, P has a local minimum between them. Either way, P' has a root between p_i and p_{i+1}.   □

To finish the proof of the Theorem, we see that we have produced $k - 1 = m + d_1 - 1$ roots of P' that differ from the de Gua points of order 1, which are also roots of P'. Thus,

$$e \geq m + d_1 - 1 + d_1,$$

i.e. $e + 1 \geq m + 2d_1$, from which we conclude

$$m \leq e + 1 - 2d_1,$$

as was to be shown.   □

4 Concluding Remarks

If nothing else, my little study of Horner's Method ought to convince any student of history of the importance of consulting primary sources whenever possible. The common declaration that Horner's Method is "practically identical"[25] to the algorithm demonstrated by Qín, which may in turn be denigrated as "merely the extension" of an earlier Chinese process, is a gross oversimplification. By all appearances, Horner brought a great deal more sophistication to the table than did Qín. He knew there could be more than one root to an equation or even none at all and incorporated this knowledge into his procedure. More importantly, he turned a slow procedure into an efficient one. When it comes to computation, efficiency and precision are what matters most and, barring some Chinese revelation, Horner wins here hands down— a victory not even hinted at in the historical accounts.

There remains, of course, the possibility that Qín or a later Chinese scholar improved the efficiency of their technique, but this is not reported by the experts and we cannot check ourselves until reliable translations of the source materials are made available.

As for Horner, the importance of having his original paper is evident when one compares its contents with Coolidge's account. Coolidge's is the most thorough I've seen and it leads one to conclude (especially with the charge of Horner's having covered his tracks) that the first paper was a tentative groping that ultimately led to his algebraic treatment, the one that I suppose is "practically identical" to Qín's version. Horner's *Transactions* paper is necessary to set that straight. The reprint in the *Ladies' Diary* is not important in understanding what Horner did, but it could partially exonerate Coolidge for having got so much so wrong.[26]

It would also be worth one's while to consult Horner's later papers to see how he treats the problem there. I fear, however, that the journals in which they reside, namely *The Mathematician* and *Leybourn's Repository,* are likely to be as rare as the *Ladies' Diary.* And, indeed, I found no copy of *Leybourn's Repository* in the state, and only one copy of *The Mathematician* in a private library in my, admittedly perfunctory, web search. A sampling of algebra texts of the 19th and early 20th centuries, probably still available in the libraries of the older universities, would constitute additional desirable primary sources— not on Horner's work, but on what would have come to be known as his method.[27] The comparison of Qín's and Horner's methods

---

[25] This is from Smith's quote given at the beginning of section 1. In *The Crest of the Peacock,* Joseph is a bit less temperate in a footnote on page 199 where he flat out states, "The procedure that Horner rediscovered is identical to the computational scheme used by the Chinese over five hundred years earlier".

[26] According to some reviews, the latest edition of Coolidge's book edited by Jeremy Gray includes an essay with many corrections. I've not seen this, but the chapter on Horner certainly merits a lot of coverage here. Of course, if the chapter is typical of the book, Gray may have been too exhausted by the time he reached that chapter to do a thorough job!

[27] In this regard, I note that in connexion with the chapter on constructibility, I consulted Julius Petersen's 1878 textbook *Theorie der algebraischen Gleichungen.* His treatment of what he calls Horner's Method presents exactly the old Chinese algorithm and not the more sophisticated approach originally taken by Horner.

becomes acceptable if by Horner's Method one means the method as taught and became common practice, if this practice indeed lacked all the bells and whistles Horner originally decked it out with.

As for my digressions on Descartes' Rule and de Gua's Theorem, I admit that I included them for mathematical rather than historical purposes, which is not to say they are historically uninteresting. I have a decent personal mathematical library, but the only proofs of Descartes' Rule I could find in any of my books were overly deep (e.g., Jacobson's classic text presents it as a corollary to Sturm's Theorem), and I thought I ought to include an elementary proof. After I worked out such I discovered both the Calculus based proof and Krishnaiah's exposition online and adapted them for inclusion here. De Gua's Theorem in its simplest form or as cited by Horner is less well-known, or, at least, it is not in my library and I thought I ought to include a proof to demonstrate the correctness of Horner's application of it. The fact that Coolidge replaced this application by his own reasoning suggests that the result was not a commonplace in the first half of the 20th century when Coolidge was active. For my discussion, I was able to download Campbell's paper via *JSTOR,* for it appeared in the *Philosophical Transactions* of the Royal Society in 1729. De Gua's proof appeared in his book published in 1740 in French. It is not listed in any of the local university libraries. Of course, knowing what he proved and how he proved it is not relevant to our underlying discussion of Horner's Method, but I would be curious to see how close I came in spirit to his proof. Come to think of it, I would be curious to see what he proved. Is the result proven here due to de Gua or to Horner or to some mysterious interpolated third figure?

Of course, what I haven't covered can be as vital a lesson for the student as what I have covered. If one had time (e.g. in a Topics in the History of Mathematics course devoted to algebra or numerical methods), one could devote some of it to a discussion of the apparent contradiction between Baron's dismissal (page 176) of Horner's claim that his method extended to transcendental functions with formula (17) which in no way requires $P(X)$ to be a polynomial. More advanced students could report on the influence of Horner's ideas in numerical analysis.

A more strictly mathematical question concerns de Gua's Theorem: what is the effect of a sequence of any odd length of 0s in the sequence of coefficients of P on the number of real roots P has? This is not history, but it could be directed into a discussion of Sturm's algorithm and its 20th century descendents (Artin-Schreier Theory, Hilbert's 17th Problem, and Tarski's Decision Procedure for Elementary Algebra and Geometry). This would bring a little 20th century mathematics into the picture and would appeal to students of algebra or mathematical logic.

However, Petersen is in the continental tradition under which, by all accounts, Horner's method was not as central as it was in the Anglo-American tradition.

8

Some Lighter Material

1 North Korea's Newton Stamps

In the May 1990 issue of Michel *Rundschau*[1], there appeared an announcement that the Peoples' Democratic Republic of North Korea would be issuing a set of 4 stamps and 1 souvenir sheet in July of that year to celebrate Sir Isaac Newton's 350th birthday.[2] Anyone familiar with North Korea's other scientific offerings– the Madame Curie, Kepler, Mendeleev, or Halley's Comet issues– looked forward to seeing their commemoration of Newton. The wait was a bit longer than anticipated, but perhaps not longer than should have been expected, as Isaac Newton's 350th birthday did not come round until 1993, when the stamps finally appeared. They were well worth the wait.

In 1993, North Korea issued not 4, but 5 stamps, and not one, but 2 souvenir sheets. While the sheets are little more than strips of 3 stamps apiece with a common, lightly decorated border, the stamps themselves are magnificent and form the finest philatelic tribute to Sir Isaac yet offered by any postal authority.

Each stamp bears the inscription "Newton. Sir Isaac/1642 - 1727". Newton's birthdate is, of course, a complicated matter. He was born on Christmas Day in 1642 according to the Julian Calendar then in force in England. According to the Gregorian Calendar, however, he was born on 4 January 1643, thus both making the inscription correct and 1993 his 350th birthyear. [Newton died on 31 March 1727 (Gregorian). The Newton stamp of the Nicaraguan Copernicus set bears the dates 1642 - 1726. This is not entirely incorrect, for, as any reader of Pepys will tell you, in those days the new year bore a double date for its first few months. Thus, March of 1727 was March of 1726/1727.]

---

[1] The supplements to the German stamp catalogue published monthly.

[2] This is a mildly edited version of an article that originally appeared in *Philatelia Chemica et Physica* 19 (1997), pp. 4 - 10, and is reproduced here with the kind permission of the editors.

The 10 Ch stamp of the set features Godfrey Kneller's portrait of 1702 of Newton at the age of 59. Kneller (1646 - 1723) was the leading portrait artist of the day and did several portraits of Newton, the first of which (also reproduced in part on a Russian stamp) is the earliest (1689) extant portrait of the scientist. Kneller's 1702 portrait, however, is the philatelically most popular and has been several times reproduced. It hangs in the National Portrait Gallery in London.

The 20 Ch stamp repeats Kneller's second portrait of Newton in cameo, but is otherwise devoted to gravitation— and a bit of biography. The Law of Gravitation is represented by the formula and an apple tree in front of Woolsthorpe Manor, Newton's birthplace.

The 30 Ch stamp again repeats the cameo. Newton's telescope, the first successful reflector and a common theme in Newton stamps, occupies centre stage. It is surrounded by the cameo and three space themes, the most relevant being a Korolev fuselage representing Newton's Third Law of Motion. There are also a radar antenna, the relevance of which escapes me, and a satellite, perhaps intended to represent orbits and Newton's derivation of Kepler's Laws of Planetary Motion???

The 50 Ch stamp is a bit of a mystery. It has the cameo and a globe with two orbiting objects, one fancy artificial satellite (perhaps the Soviet space station) and a dot (a more primitive satellite no doubt). The centre-piece is a formula— the Finite Binomial Theorem. This is an unfortunate

error. Newton's contribution to the Binomial Theorem was the infinite series representation for exponents other than positive integers; the form presented was known already to Blaise Pascal some years earlier. The notation is also certainly not Newtonian. The symbol C_n^k is the familiar binomial coefficient, which is usually written $\binom{n}{k}$ and occasionally $_nC_k$ or C_k^n. Thus, the notation is slightly incorrect.

The real mystery of the fourth stamp is not what the space symbols represent or how the super- and sub-scripts in the binomial coefficients got switched, but what is lurking in the background. Hiding behind the Binomial Expansion are a geometric diagram that I couldn't find in my copy of the *Principia* and a handwriting sample I similarly could not identify.

The 70 Ch stamp does not bear Kneller's portrait in cameo. Instead it features the statue of Newton from his tomb at Westminster Abbey, minus some of its kitschier surroundings. The statue deserves some comment. It depicts Newton reclining on a stack of 4 books, his *Divinity*, *Chronology*, *Opticks*, and *Principia*. It is not clear from the reproduction on the stamp just how many books there are, or, indeed, that his support consists of books. Westfall's description[3] of the whole reveals not only what is left out of the philatelic reproduction, but why it was left out. It is, according to the famous Newton scholar, "a baroque monstrosity with cherubs holding emblems of Newton's discoveries, Newton himself in a reclining posture, and a female figure representing Astronomy the Queen of the Sciences, sitting and weeping on a globe that surmounts the whole. Twentieth-century taste runs along simpler lines, and the monument is now roped off in the Abbey so that one can scarcely even see it." Maria Mitchell, America's first woman astronomer, has also left us a description of the statue[4]:

> The base of Newton's monument is of white marble, a solid mass large enough to support a coffin; upon that a sarcophagus rests. The remains are not enclosed within. As I stepped aside I found I had been standing upon a slab marked 'Isaac Newton', beneath which the great man's remains lie. On the side of the sarcophagus is a white marble slab, with figures in bas-relief. One of these imaginary beings appears to be weighing the planets on a steel-yard. They hang like peas! Another has a pair of bellows and is blowing a fire. A third is tending a plant.
>
> On this sarcophagus reclines a figure of Newton, of full size. He leans his right arm upon four thick volumes, probably 'The Principia', and he points his left hand to a globe above his head on which the goddess Urania sits; she leans upon another large book.
>
> Newton's head is very fine, and is probably a portrait. The left hand,

[3] Richard S. Westfall, *Never at Rest; A Biography of Isaac Newton*, Cambridge University Press, Cambridge, 1980.

[4] Phebe Mitchell Kendall, *Maria Mitchell; Life, Letters, and Journals*, Lee and Shepard Publishers, Boston, 1896.

which is raised, has lost two fingers. I thought at first that this had been the work of some 'undevout astronomer'[5], but when I came to 'read up' I found that at one time soldiers were quartered in the abbey, and probably one of them wanted a finger with which to crowd the tobacco into his pipe, and so broke off one.

The missing fingers offer an alternative to Westfall's Art-Critic Theory of the roping off of the monument.

The elements of the background of the 70 Ch stamp are these: A geometrical diagram from the proof of Proposition LXXI, Theorem XXXI, of the *Principia* is fairly faithfully reproduced: a capital "I" has been changed into a lower case "t". The proposition in question asserts that a particle external to a homogeneous spherical body can be taken to be attracted to the centre of the sphere. Behind this diagram are the title pages from the first editions of Newton's works. Clearly discernible are those of the *Opticks* (1687) and *Principia* (1704). Behind these are more title pages, and it is entirely possible that a Newton scholar can identify the two sticking out from behind the *Principia* by the single letters visible on them.

The sheets, as I said, offer no great additions. Each contains the 10 Ch stamp bearing Kneller's portrait. One also includes the 20 and 70 Ch stamps,

---

[5] "An undevout astronomer is mad". This is a popular quotation from a poem by Edward Young (1683 - 1765).

while the other has the 30 and 50 Ch ones. The identical borders, in blue, have various line drawings, mostly devoted to space travel. The exceptions are another drawing of Newton's telescope, Woolsthorpe Manor seen from another perspective, and the title page of *Principia*.

2 A Poetic History of Science

GREAT WORKS OF MAN[6]

Imhotep was first who designed
A tomb of pyramidal kind.
Thus did the physician
Begin a tradition
Of monuments likewise outlined.

IRRATIONALITY

Pythagoras woke from his slumbers.
He'd dreamt that all things were numbers.
But alas he soon knew
That the square root of two
His theory with paradox encumbers.

DISCOVERY OF THE LEVER

A Greek, the famous Archimede',
Said, "Progress you cannot impede.
Though the Earth may seem still
I can move it at will;
A fulcrum is all that I need".

ATTACKED FROM BEHIND

Avicenna was poisoned my friend
Through a physic received in the end.
Now broken bones tend
Like fractures to mend;
So its better to break than to bend.

---

[6] I take pleasure in acknowledging my indebtedness to my poetic mentor Lydia Rivlin, who not only ripped apart my primitive attempts at high poetry, but also patiently explained scansion and even contributed some improved lines. In one case, she rewrote the entire poem, but I still liked my earlier version and include all the variants here.

[Alternate Version

Avicenna was purged in the end.
From its poison he never did mend;
Yet bones can be set
By every vet.
So its better to break than to bend.]

[Lydia's Version

That Avicenna was poisoned is known
From what's written upon his gravestone.
Although accidental[7]
It is quite fundamental:
If you poison your bottom, you're blown.]

ABSENT FROM THE CREATION

For hubris no doubt the first prize
Belongs to Alfonso the Wise,
Whose lips when unfettered
Said HE could have bettered
God's plan for the stars in the skies.

FIBONACCI NUMBERS

Fibonacci ran after a rabbit.
He shouted, "Oh, please help me grab it".
As it ran away
We all heard him say,
"The count's not complete till I nab it".

SCIENCE IN THE THIRD WORLD

Qín now sits in the corner–
A footnote to Dubbya G. Horner.
It really don't matter
He beat out the latter:
We ain't gonna credit no for'ner[8].

---

[7] According to his autobiography, completed by one of his students, Avicenna was definitely murdered. Some scholars, however, believe the poisoning to have been accidental.

[8] Poetic license allows me to spell "foreigner" this way.

Paracelsus

Declaring the Greek was now past us,
The fiery Doctor Bombastus
Put Galen to flame
In Progress's name—
The faculty thought this prepost'rous.

Elizabethan Tragedy

John Dee was a man who had yearned
For knowledge that's best left unlearned.
So the masses all came
And in God's holy name,
His house and his books were thus burned.

Calculus

I hear the two Brothers Bernoulli
Were more than a trifle unruly.
The day even came
When each would exclaim
The other was famous unduly.

Non-Euclidean Geometry

A pity that Carl F. Gauss
Young Bolyai's ambitions did douse.
They shared the lad's claim
To parallel fame
In tearing down Euclid's old house.

Statistical Error

You heard what the rabble all say
About the Marquis Condorcet?
He was caught my dear cousin
Eating eggs by the dozen—
His appetite gave him away.

Survival of the Fittest

Completing long years of reflexion
On his biologic collection
Old Darwin resolved
That species evolved,
Directed by nat'ral selection.

INVENTION OF THE PIE CHART

Miss Nightingale, famed for her lamp,
Compiled the statistics in camp.
With her chart like a pie
All was clear to the eye;
Thus did medical practice revamp.

MODERN COMPUTING

A computer designed by Babbage
Got instructions on cards via stabbage.
But nowadays dudes
Can input their nudes
Through devices for pictorial grabbage.

ASTRONOMER ROYAL

George Biddel was really quite wary.
For Neptune he deigned not to tarry.
The dateline he flouted;
'Gainst Babbage he spouted.
Inside of his head 'twas all Airy!

PSYCHOPATHOLOGY

The theories of Sigmund of Freud
The masses of people annoyed.
They found it no joy
To learn that each boy
With thoughts of his mother had toyed.

SETS AND VIOLENCE

"All integers come straight from God,"
Said Kronecker with a big nod.
But Cantor he trounced,
As his Sets were denounced.
Which drove the poor fellow quite odd.

TOPOLOGICAL PUZZLE

And here is a bottle of Klein,
A wonder of modern design.
But a question about
Its mysterious spout—
Who knows how to pour out its wine?

HISTORIOGRAPHY I: THE MAKING OF A LEGEND

Old Cantor's hypothesis sketched
A picture— Egyptian ropes stretched,
A square to derive
Through lengths 3, 4, 5.
The story in stone was soon etched.

HISTORIOGRAPHY II: POTBOILERS

Now Eric the Templer named Bell
Had quite a good story to tell.
He may have been lax
In checking his fax[9],
But truth's not what made his work sell!

DISCOVERY OF INSULIN

Remember when Frederick Banting
Indulged in some raving and ranting?
But put to the test
He needed the Best
Of the aid that Macleod had been granting.

LINDEMANN REDIVIVUS

Indiana through legal contortion
One day tried to put π in proportion.
But the bill in its path
Met a teacher of math
Who arranged for its timely abortion.

DEUTSCHE PHYSIK

When Einstein discovered a cavity
In Newton's old theory of gravity,
No apples of red
Came down on his head;
But only a charge of depravity.

---

[9] Since I don't deal with fax machines, I take it for granted that everyone will know I mean "facts". I chalk this up to poetic license, but fear it may actually be a sign that I am antiquated in that my work requires annotation.

Banach-Tarski Paradox

Both Banach and Tarski were Poles
With anti-euclidean goals.
They proved, so I hear,
The parts of a sphere
To equal a couple of wholes.

Nonconstructive Criticism

As Hilbert exhorted to Brouwer
When matters between them went sour,
Experimentation
With double negation
Gives mathematicians their power.

Spectre

Old Brouwer was turned into toast.
But Hilbert was really engrossed
In life and in dreams,
At least so it seems,
In battle with Kronecker's ghost.

Proof Theory

I doubt you'll find any dents in
The theories of Gerhard of Gentzen.
For he had a mind
Of orderly kind,
Arranging his thoughts in *Sequenzen*.

Deep Thoughts I

With Beebe[10] and Barton as crew
The bathysphere sank out of view.
Where others would drown
A half-a-mile down
They found an exotic new zoo.

---

[10] Pronounced BEE-bee.

DEEP THOUGHTS II

Its quite true that August Piccard[11]
Went deeper than Beebe by far.
But each of the pair
In watery lair
Saw things that were truly bizarre.

HIGH HOPES

Gagarin was launched into space,
First victory claimed in the race −
A victory hollow,
For soon with Apollo
America landed first place.

FERMAT'S LAST THEOREM

Old Fermat's final problem beguiles,
Its attackers soon losing their smiles.
Though so simple to state,
Its solution of late
Has required all our wits and our Wiles.

3 Drinking Songs

From the 8th to the 13th of August, 1904, the Third International Congress of
Mathematicians was held in Heidelberg. In conjunction with this event B.G.
Teubner Verlag issued a slim little volume of beer drinking songs under the
lengthy title,

Liederbuch
den Teilnehmern am
Dritten Internationalen
Mathematiker-Kongress
in Heidelberg
als Andenken
an die Tage vom 8. vis 13. August 1904
überreicht von der
Deutschen Mathematiker-Vereinigung[12]

---

[11] The "d" is silent.

[12] In English: *Songbook for the Participants in the Third International Congress of
Mathematicians in Heidelberg as a Memento of the Days from 8 to 13 August
1904 Presented by the German Mathematicians Union.*

The collection is divided into three parts— general songs (numbers 1 to 22), mathematical songs (numbers 23 to 33), and some German folk songs (numbered anew from 1 to 6). Of interest here are the mathematical songs:

23. "Gauss zum Gedächtnis" ("Gauss in Remembrance"), by Eugen Netto

24. "Alte und neue Zeit" ("Olden Times and New"), by Hermann Schubert

25. "International, Ein Lied in 14 Sprachen" ("International, A Song in 14 Languages"), by Netto, Schubert and Companions

26. "Die Rückkehrpunkte" ("The Turning Point"), by Moritz Cantor

27. "Pythagoras", by Moritz Cantor

28. "Unser guter Raum" ("Our Good Space"), by Hermann Schubert with the support of Kurt Lasswitz[13]

29. "Die moderne Richtung" ("The Modern Direction"), by Eugen Netto

30. "Das Rendez-vous der Parallelen" ("The Rendezvous of the Parallels"), by Moritz Cantor

31. "Die unglückliche Liebe" ("Unrequited Love"), by Hermann Schubert

32. "Bierlied" ("Beer Song"), by Eugen Netto

33. "Popularisierung der Mathematik" ("Popularisation of Mathematics"), by Hermann Schubert.

The mathematical songs are to be sung to the tunes of various well-known beer drinking songs. The music is not included, but the original lyrics sometimes are. For example, number 24 "Alte und neue Zeit" is to be sung to the tune of "O alte Burschenherrlichkeit" ("O, Good Old Student Days"), a popular and sentimental tune reproduced as song 9 in the book. "International"

---

[13] Lasswitz, whose name can also be rendered Kurd Lasswitz or Kurd Laßwitz, is responsible for another mathematical poem. His "Prost, der Faust Tragödie (n)-ter Teil" of 1882 can be found in Waltraud Wende-Hohenberger and Karl Riha, eds., *Faust Parodien*, Insel Verlag, Frankfurt am Main, 1989. It begins with the student Prost having difficulty integrating a difficult differential equation. Lasswitz, the "Father of German Science Fiction", was a writer as well as a mathematician. A sample of his mathematical fiction more accessible to American readers can be found in Clifton Fadiman's anthology, *Fantasia Mathematica*, Simon and Schuster, New York, 1958. This latter book, incidentally, is another source for several good mathematical poems, including two limericks by Cyril Kornbluth and Sir Arthur Eddington that I wish I had been clever enough to have written.

is to be sung to the tune of "Im schwarzen Wallfisch zu Askalon" ("Inside the Black Whale of Askalon"), song number 17. "International" has but one stanza of 4 lines repeated 14 times in 14 different languages from Greek to Volapük. The English verse reads:

> In ancient times, upon the door
> Of Plato, there was writt'n:
> "To each non-mathematicus
> The entrance is forbidd'n."

The material of the preceding section notwithstanding, I declare myself insufficiently poetically gifted to attempt a poetic translation of these songs, one that would respect the rhyme and rhythm while still being faithful to the contents. What I can do is present one of the songs in its original musical German and accompany it by a loose line-by-line English translation. This should give one something of the flavour of the collection. I have chosen the following song by Eugen Netto.

Die moderne Richtung	*The Modern Direction*
Ich weiss nicht, was soll es bedeuten,	I don't know what it all means
Dass ich so traurig bin!	That I'm so very sad!
Die alten, die seligen Zeiten,	The old, the blissful times
Die sind nun auf ewig dahin.	They are now forever gone.
Des Zweifel's Wogen schwellen	Waves of doubt are swelling
Im wildbewegten Meer;	In fiercely moving sea;
Was man vor Jahren wusste,	What one knew in previous years,
Das glaubt man heut' nicht mehr.	One believes no more today.
Ja früher verliefen Funktionen	Yea, functions proceeded
Gemächlich, in stetigen Schritt;	Leisurely, in continuous pace;
Die wenigen Unendlichkeits-Punkte,	The few infinity-points,
Die zählten wahrhaftig kaum mit.	One hardly bothered to count.
Natürlich war jegliche Kurve	Naturally any curve was
Mit Richtung und Krümmung verseh'n;	Seen with direction and curvature;
Und was "Dimension" sei, das konnte	And what "dimension" was, could
Jedweder Sextaner versteh'n.	Every freshman comprehend.
Doch heute? Dreifaches Wehe,	But today? Triple woes,
Wie alles sich jetzt kompliziert!	As everything more complex grows!
In Anseh'n steht nur die Kurve,	In sight are only the curves
Die unendlich oft oszilliert.	Which oscillate infinitely often.
Ableitungen sind aus der Mode,	Derivatives are out of fashion,
Tangenten fehlen total.	Tangents fail totally,
Singularitäten erblühen	Singularities blossom
Im allertollsten plural.	In maddening profusion.

Den Stud. math. in höh'rem Semester	The math student in later semester
Ergreift es mit wildem Weh',	Grasps it with wild grief.
Er brütet bei Nacht und bei Tage	He broods by night and by day
Und hat doch keine Idee.	And still hasn't a clue.
Ich glaube, er fällt durch's Examen,	I believe, he takes his exams
Mit Schrecken schon sieht er es nah'n:—	With Terror he sees it coming:—
Das hat die moderne Richtung	This has the modern direction
Der Math'matik ihm getan.	Of Mathematics done to him.

The theme of the change of mathematical direction as a change for the worse was not unique to this song. Hermann Schubert's "Alte und neue Zeit" also touches on the subject. It would reach an extreme in the 1930s with the nazi distinction between good German-Aryan *anschauliche* (intuitive) mathematics and the awful Jewish tendency toward abstraction and casuistry.[14] In 1904, however, the mood was cheerful and the complaint about the new direction was imbued with an evident tongue-in-cheek humour and wistful nostalgia, and was not a call-to-arms.

An English wine drinking song, more than half a century older than the Heidelberg repertoire, is reported on by Augustus de Morgan in his *Budget of Paradoxes*[15]. The song was written on the occasion of a party honouring a member of the Mathematical Society (1767 – 1845) who was also a solicitor and who had defended the Society in some legal action. When the Society's membership had shrunk to a non-sustainable size, it was absorbed into the Astronomical Society and the song fell into the latter's possession. Eventually, it fell into de Morgan's hands and he published it as follows. I include a few of his footnoted remarks by way of annotation.

THE ASTRONOMER'S DRINKING SONG

Whoe'er would search the starry sky,
   Its secrets to divine sir,
Should take his glass— I mean, should try
   A glass or two of wine, sir!
True virtue lies in golden mean,
   And man must wet his clay, sir;
Join these two maxims, and 'tis seen
   He should drink his bottle a day, sir!

Old Archimedes, reverend sage!
   By trump of fame renowned, sir,

---

[14] The proponents of this distinction had to dance some fancy steps in explaining how the abstract mathematics of David Hilbert, the foremost German mathematician of the previous quarter century, was not the bad abstraction of the Jews. A particularly vitriolic, yet revealing, example of an attack on Jewish mathematics by Hugo Dingler is quoted in Eckart Menzler-Trott, *Logic's Lost Genius: The Life of Gerhard Gentzen*, American Mathematical Society, to appear.

[15] Cf. the Bibliography.

Deep problems solved in every page,
  And the sphere's curved surface found, sir:
Himself he would have far outshone,
  And borne a wider sway, sir,
Had he our modern secret known,
  And drank a bottle a day, sir!

When Ptolemy, now long ago,
  Believed the earth stood still, sir,
He never would have blundered so,
  Had he but drunk his fill, sir:
He'd then have felt it circulate,
  And would have learnt to say, sir,
The true way to investigate
  Is to drink your bottle a day, sir!

Copernicus, that learned wight,
  The glory of his nation.
With draughts of wine refreshed his sight,
  And saw he earth's rotation;
Each planet then its orb described,
  The moon got under way, sir;
These truths from nature he imbibed
  For he drank his bottle a day, sir!

The noble[16] Tycho placed the stars,
  Each in its due location;
He lost his nose[17] by spite of Mars,
  But that was no privation:
Had he but lost his mouth, I grant
  He would have felt dismay, sir,
Bless you! *he* knew what he should want
  To drink his bottle a day sir!

Cold water makes no lucky hits;
  On mysteries the head runs:
Small drink let Kepler time his wits
  On the regular polyhedrons:
He took to wine, and it changed the chime,
  His genius swept away, sir,

---

[16] The common epithet of rank: *nobilis Tycho*, as he was a nobleman. The writer
had been at history

[17] He lost it in a duel with Manderupius Pasbergius. A contemporary, T.B. Laurus,
insinuates that they fought to settle which was the best mathematician! This
seems odd, but it must be remembered they fought in the dark, "*in tenebris
densis*"; and it is a nice problem to shave off a nose in the dark, without any
other harm.

Through area varying[18] as the time
  At the rate a bottle a day, sir!

Poor Galileo, forced to rat
  Before the inquisition,
E pur si muove was the pat
  He gave them in addition:
He meant, whate'er you think you prove,
  The earth must go its way, sirs;
Spite of your teeth I'll make it move,
  For I'll drink my bottle a day, sirs!

Great Newton, who was never beat
  Whatever fools may think, sir;
Though sometimes he forgot to eat,
  He never forgot to drink, sir:
Descartes[19] took nought but lemonade,
  To conquer him was play, sir;
The first advance that Newton made
  Was to drink his bottle a day, sir!

D'Alembert, Euler, and Clairaut,
  Though they increased our store, sir,
Much further had been seen to go
  Had they tippled a little more, sir!
Lagrange gets mellow with Laplace,
  And both are wont to say sir,
The *philosophe* who's not an ass
  Will drink his bottle a day, sir!

Astronomers! What can avail
  Those who calumniate us;
Experiment can never fail
  With such an apparatus:
Let him who'd have his merits known
  Remember what I say, sir;
Fair science shines on him alone
  Who drinks his bottle a day, sir!

How light we reck of those who mock
  By this we'll make to appear, sir,
We'll dine by the sidereal clock
  For one more bottle a year, sir,

---

[18] Referring to Kepler's celebrated law of planetary motion. He had previously
wasted his time on analogies between the planetary orbits and the polyhedrons.

[19] As great a lie as ever was told: but in 1800 a compliment to Newton without a
fling at Descartes would have been held a lopsided structure.

But choose which pendulum you will,
    You'll never make your way, sir,
Unless you drink— and drink your fill,—
    At least a bottle a day, sir!

4 Concluding Remarks

Of the material presented in this chapter, the only item of serious historical import is Netto's beer drinking song which, even in my bad translation, bears eloquent witness to a sea change in mathematics, a turn in direction away from intuition and towards formalism. This trend, begun in the mid-19th century with the arithmetisation of analysis would reach newer and newer heights of abstraction as the 20th century wore on. The astronomer's drinking song offers little of history other than a list of names of those individuals in the history of astronomy deemed important at the time of the song's composition. My own limericks cannot be said to pass a similar judgment, for, as the author I know first hand that considerations of rhyme and punchline outweighed the importance of an individual's particular contributions to science in deciding whom to include. Chance also played a rôle: a late addition was the stanza on Horner, hardly a central figure in the history of mathematics, but one who was on my mind as I set about collecting material for the present chapter after finishing writing Chapter 7.

My essay on the North Korean Newton stamps will not teach the reader anything about Newton or the history of mathematics he couldn't learn elsewhere. But it does illustrate one way in which stamp collecting has been educational for me. Setting about identifying the various elements incorporated in the stamps' designs was a focussed effort— and focus always makes the results of one's research more memorable.

I have used the results of my stamp collecting elsewhere in this book. There are several thousand scientists, physicians, and technologists appearing on stamps. To keep track of them I maintain a small database, part of which is a list of 3310 individuals with variant spellings of their names, along with variant birth and death dates as given in several standard references including the *Dictionary of Scientific Biography*, the *Encyclopædia Britannica*, the *World Who's Who of Science*, the *Brockhaus Enzyklopädie*, and the *Great Soviet Encyclopedia*. My little exercise in Chapter 2, section 1, on Camille Flammarion and Carl Auer von Welsbach can be traced back to this database. Other uses of such a database are i) compiling a list of famous scientists who were born or died on one's birthday; ii) having the ability to announce in class that "Today is the birthday of _____ who _____"; and iii) making a quick assessment of the reliability of a general reference work. Birth and death dates may be the least important details in the history of science, but a reference that gets these minor details correct inspires more confidence than one that doesn't. Using near unanimity of agreement to determine "correct" birth and death dates, one can count the number of errors the various reference works have made in one's sampling of individuals to measure the relative reliability or unreliability of the works. Not surprisingly, the *Dictionary of Scientific Biography* tops my list of reliable sources.

Many is the scientist who is not on my list. Like the list of subjects of my poems, the list of scientists on stamps is not necessarily representative. Figures who should be honoured by their countries and their importance proclaimed to the world may have the misfortunate of having been citizens of countries like the United States and England which almost make a point of not depicting scientists on stamps[20], or they may have the misfortune of having lived under the shadow of a far more famous scientist who hogs all the philatelic glory his nation has to offer[21]. At the opposite extreme, a comparatively unimportant scientist may merit philatelic recognition by a small country because of a local connexion. For example, the marine zoologist Joseph Jackson Lister is sufficiently obscure as not to be in any of the standard reference works other than the Spanish encyclopædia *Ilustrada*. His voyages in the South Pacific, however, must be regionally famous as his portrait graces a stamp issued by Christmas Island in 1978. Albert Einstein's first wife Mileva Marić was Serbian and ap-

---

[20] As I write these words, the United States has recently broken this longstanding tradition by issuing stamps honouring Barbara McClintock (Nobel prize winning geneticist), Josiah Willard Gibbs (mathematician and physicist, one of the founders of vector analysis), John von Neumann (one of the most famous mathematicians of the 20th century and one of the few known to the general public), and Richard Feynman (Nobel prize winning physicist who became a celebrity shortly before his death). Einstein was so famous he was honoured twice by the United States Postal Service, but Theodore von Karman whose contributions to the country were vastly greater required years of petitioning to be commemorated on an American stamp. Generally speaking, in recent decades the only scientists to appear on American stamps were in the Black Heritage series, and none of them were mathematicians, unless you count Benjamin Banneker who should more correctly be termed a mathematical practitioner, for he made no contributions to the field itself.

[21] So long as Abel is alive— er, I mean dead— I have no hope of seeing Thoralf Skolem or Axel Thue on Norwegian postage stamps. And Sweden annually issues 3 or so stamps of Nobel prize winners, very often in the science categories. So I don't expect to see Gösta Mittag-Leffler on a Swedish stamp any time in the near future. [A possibly amusing aside: My point here was simply that, with its budget of science stamps taken up by the Nobel prize, one shouldn't expect to see any mathematicians on Swedish stamps— not even its most famous mathematician Mittag-Leffler. The referee misunderstood my point and wrote, "Some readers will not know the story of Nobel and Mittag-Leffler; in any case, I believe the story of the 'mistress' has been debunked. I had actually forgotten that American oral tradition had it that there is no Nobel Prize in mathematics because Nobel learned that Mittag-Leffler had been fooling around with Nobel's mistress. This story never had greater authority than that of an urban legend, it betrays complete ignorance of Nobel's intention with his prize, and it doesn't need debunking.]

peared on a Serbian *Europa* [22] stamp in 1996. The justification for this stamp was purely ethnic pride. For, the theoretical justification for her fame, other than having been married to Einstein, was a book by a retired middle school mathematics teacher claiming, without any supporting evidence, that Marić was responsible for Einstein's special theory of relativity. Then too, someone of genuine but not major mathematical interest could well be celebrated on stamps for nonmathematical reasons— Jan de Witt, author of an early algebra book, was also mayor of Amsterdam; Louis Antoine de Bougainville, who wrote a calculus text, is mostly remembered for circumnavigating the world; and, of course, Napoleon Bonaparte is unlikely ever to be remembered on a stamp for his little theorem on triangles.

Stamps featuring mathematicians and scientists may, as with the North Korean Newtons, represent iconically the life and work of the individuals. In such a case, it may be rewarding to hunt down the meanings of the various icons. England's 1991 stamp honouring Charles Babbage, however, offers an example where the hunt will not prove rewarding: whatever an empty head filled with numbers is supposed to represent I don't think even the artist knows. In one case, the symbolism was incomplete: when Austria issued a stamp honouring the philosopher Ludwig Wittgenstein in 1989, it included a cameo of an owl, but left out the mirror.

Often stamps honouring mathematicians feature only portraits. Obviously, portraits of anyone too ancient are mere guesswork, or symbolic reconstruction— a large brow to denote intelligence or a white beard to signify age and the wisdom that supposedly comes with it. Rembrandt's painting of Aristotle contemplating the bust of Homer depicting Aristotle in Renaissance attire appears on a stamp issued by the Grenadines of Grenada in 1993, and Socrates appears in Arab dress on a stamp issued by Ajman in 1967. Even greater license was taken by the French in 1964 when they issued a stamp featuring a portrait of the mathematician Gerbert[23], also known as Pope Sylvestre II, a man of whom no authentic likeness exists. With more modern

---

[22] Each year the European nations each issue a pair of stamps devoted to a single topic. The stamps of this series are called *Europa* stamps. In 1996 the topic was Famous Women.

[23] And subject of a short poem by Walther von der Vogelweide. Gerbert's use of arabic numerals was deemed by many a sort of black magic— his "familiar" was reputedly named Abacus. He was believed by many, especially the later Protestants, to have been damned for his infernal dealings. Thus von der Vogelweide wrote,

> The chair at Rome is now properly filled,
> as it was formerly by the magician Gerbert.
> He plunged into ruin only his own one soul:
> the present one will ruin himself and all Christendom.

The rhyming is better in the original German.

individuals there are usually genuine portraits or even photographs and the depictions on stamps are fairly accurate. There is one glaring exception.

As early as 1951, L. von David reported that János Bolyai, one of the founders of non-Euclidean geometry, took a sabre to the only existing portrait of himself and, consequently, he bequeathed no image of himself to posterity. Notwithstanding this, in 1960, on the occasion of the centenary of his death, both Hungary and Romania issued stamps bearing his supposed portrait.[24] This common portrait was immediately accepted as genuine and found its way into a number of history books and general reference works. I reproduce the Hungarian stamp on the right.

In my own library I find the portrait represented as authentic in the following works.

- Hans Reichardt, *Gauß und die Anfänge der nicht-euklidischen Geometrie* (*Gauss and the Origins of Non-Euclidean Geometry*), B.G. Teubner, Leipzig, 1976.[25]

- W. Gellert, H. Küstner, M. Hellwich, and H. Kästner, eds., *The VNR Concise Encyclopedia of Mathematics*, Van Nostrand Reinhold Company, New York, 1977.

- Hans Wussing and H. Arnold, eds., *Biographien bedeutender Mathematiker* (*Biographies of Significant Mathematicians*), Aulis Verlag Deubner and Co. KG, Köln, 1978.

---

[24] Bolyai was Hungarian. However, today both his birthplace and place of death are in Romania and both countries claim him as their own. To me this is a familiar affair: my mother's father was Hungarian, yet when I saw some years ago one of his official documents (perhaps a copy of his birth certificate) it had a Romanian stamp affixed.

[25] Actually, I have the 1985 reprint which includes an appendicial acknowledgment that the portrait is not of Bolyai, but of some unknown contemporary. Incidentally, in another appendix, a scene from Kurt Lasswitz's *Faust* parody cited in footnote 13 is reproduced.

- Victor Katz, *A History of Mathematics; An Introduction*, Harper Collins, New York, 1993[26].

- Robin J. Wilson, *Stamping Through Mathematics*, Springer-Verlag, New York, 2001.

Not every author in my library was fooled. In the preface to his 1974 *Euclidean and Non-Euclidean Geometries; Development and History*[27], Marvin J. Greenberg explains the absence of a portrait of János Bolyai to accompany all the other portraits in his book in expressing his gratitude "to Professor István Fáry... for contacting Hungarian mathematicians in an attempt to obtain a portrait of János Bolyai (apparently no authentic portrait exists)". Peter Schreiber noted without explanation in his book on mathematical stamps[28] that the portrait on the stamp was not of Bolyai, but some unknown contemporary. In the 1985 reprint of his book on non-Euclidean geometry, Reichardt reports that the editors at Teubner contacted the Romanian and Hungarian postal authorities and received acknowledgment that the portrait was not authentic. Hans Wussing, mathematician, historian, and philatelist, and co-author Horst Remane included an illustration of the Hungarian stamp in their book *Wissenschaftsgeschichte en miniature*, but noted too that there is no authentic portrait of Bolyai. Wussing also included an image of one of the stamps and a note, with no explanation, that the stamp does not bear an authentic portrait, in an article on philately in the second volume of Ivor Grattan-Guinness's *Companion Encyclopedia in the History and Philosophy of the Mathematical Sciences*, cited in section 13 of the Annotated Bibliography.

The portrait is widespread and the statement of its lack of authenticity less so. I haven't come across the portrait in any paper publication later than the 1998 edition of Katz's book. However, as recently as the night before these words were being written the bogus portrait could be found on the MacTutor History of Mathematics website of the School of Mathematics and Statistics of the University of St. Andrews in Scotland, albeit with the warning at the bottom of the page that the portrait may not be authentic. Further web surfing

---

[26] A visit to one of the better local municipal libraries revealed that the bogus portrait was retained in the second edition published in 1998.

[27] W.H. Freeman and Company, San Francisco

[28] See section 14 of the Annotated Bibliography for the full reference as well as for references to other mathematico-philatelical works.

revealed that Hungarian mathematicians have created some virtual portraits of János Bolyai by morphing existing portraits of his father and his son[29][30].

The episode of the Bolyai stamps is a bit atypical. One might expect to see stamps used as illustrations or even as objects of mini-research projects as students seek to explain the elements of their designs; one does not expect them to be the subjects of history themselves or to rewrite history as the Bolyai stamps are and did. Well, maybe I'm exaggerating here, but a couple of generations of mathematicians have had that image of the adolescent Bolyai burned into their brains alongside the more staid and mature portraits of Nikolai Ivanovich Lobachevsky and Carl Friedrich Gauss, and I wouldn't be surprised if it had its effect.

Postage stamps, poems, more general works of fiction— these generally will not tell us much about the history of mathematics directly, but they can lighten up the course and offer mini-modules that can be exported to other math courses. Being more serious myself, I shall end on a high note, leaving roughly where we came in with the following exercise:

Assignment. During the American folk music revival of the 1960s, singers Bob Gibson and Bob Camp sang an updated version of the classic "John Henry" in which the "steel drivin' man" became a "thinkin' man" and challenged a computer instead of the traditional steam drill. Among the lyrics we hear

> Now the man who invented the computer
> Was from a place called M.I.T.
> He punched out cards and tapes by the yards
> Humming, "Nearer My God to Thee".

Whom were they singing about and how accurate was their assessment of where the credit was due?

---

[29] I must watch too many crime shows on television, for it strikes me the correct approach is to exhume the body and turn the skull over to a forensic anthropologist to apply layers of clay to. Fortunately, my taste in science fiction is for the classic films of the 1950s, so I won't suggest extracting DNA from the remains and cloning him.

[30] A new book by Jeremy Gray entitled *János Bolyai, Non-Euclidean Geometry, and the Nature of Space* and published by MIT Press (Cambridge, Mass.) has just been listed on Amazon.com. The reproduction there of the table of contents lists a short, 2-page chapter, "A portrait at last". Presumably this is the Hungarian morph.

A

Small Projects

1 Dihedral Angles

The determination of the dihedral angles given on page 82 is involved, but conceptually simple. The easiest case is that of the octahedron. We can think of the octahedron as two four-sided pyramids glued together at their square bases. Half the dihedral angle would be given by the angle between one of the faces and the base. Without loss of generality, we can assume the edges to be of length 1 and the base to lie in the xy-plane with its vertices at $(0,0)$, $(1,0)$, $(1,1)$, and $(0,1)$:

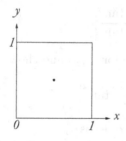

Figure 1

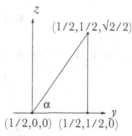

Figure 2

The apex of the pyramid lies directly above the centre of the square and thus has coordinates $(1/2,1/2,z)$ in xyz-space for some value of z. To determine z, note the distances of $(1/2,1/2,z)$ to $(0,0,0)$, $(1,0,0)$, $(1,1,0)$, and $(0,1,0)$ must all equal

$$\left(\frac{1}{2}-0\right)^2 + \left(\frac{1}{2}-0\right)^2 + (z-0)^2 = \frac{1}{4} + \frac{1}{4} + z^2 = 1,$$

whence $z^2 = 1/2$ and $z = \pm\sqrt{2}/2$. For the vertex above the plane, this yields $z = \sqrt{2}/2$.

The plane $x = 1/2$ contains the apex and its perpendiculars to one of the faces and the xy-plane. The angle α between this face and the base satisfies

$$\tan \alpha = \frac{\sqrt{2}/2}{1/2}.$$

(Cf. *Figure 2.*) Thus, $\alpha = \tan^{-1}(\sqrt{2}) = 54.7356°$. Doubling this yields $109.47°$.

Repeat this to find the dihedral angles of the tetrahedron, dodecahedron, and icosahedron.

2 Inscribing Circles in Right Triangles

Consider the following variant of Problem 2.9 of Chapter 4. Given a length a and a radius r, construct a right triangle with hypotenuse a and an inscribed circle of radius r.

As in 2.11 - 2.13 of Chapter 4, show that there is a constant C such that, if $a \geq Cr$, there is a right triangle of hypotenuse A with an inscribed circle of radius r. [Hint: Imagine the right triangle positioned in the xy-plane in such a way that the endpoints of the hypotenuse are $(0,0)$ and $(a,0)$, with the third vertex in the first quadrant. Let α be one of the acute angles of the triangle and $\beta = \pi/2 - \alpha$ the other angle. The centre (x,y) of the inscribed circle lies on the bisectors of these angles. Show

$$\tan \frac{\beta}{2} = \frac{1 - \tan \frac{\alpha}{2}}{1 + \tan \frac{\alpha}{2}},$$

use the equations of the bisectors to solve for (x,y) and show

$$r = a\frac{1 - \tan \frac{\alpha}{2}}{1 + (\tan \frac{\alpha}{2})^2} \cdot \tan \frac{\alpha}{2}$$

$$= \frac{a}{2} \cdot (-1 + \sin \alpha + \cos \alpha).$$

Conclude $C = 2\sqrt{2} + 2$.]

If $a \geq Cr$ and a, r are constructible, does it follow that the right triangle in question can be constructed by ruler and compass?

3 cos 9°

Equation (19) of Chapter 4 yields the equation

$$32X^5 - 40X^3 + 10X - \sqrt{2} = 0 \tag{1}$$

for $\cos 9°$.

i. Substitute $Z = \sqrt{2}X$ into (1) to get

$$4Z^5 - 10Z^3 + 5Z - 1 = 0.$$

Divide by $Z + 1$ to obtain a 4th degree equation satisfied by $\sqrt{2}\cos 9$. Look up the solution to the general 4th degree equation in a standard History of Mathematics textbook and apply it to obtain an exact expression for $\cos 9$.

ii. Simplify

$$\left(32X^5 - 40X^3 + 10X\right)^2 - 2 = 0$$

to obtain a 10th degree equation with integral coefficients. Apply Horner's method to this to approximate $\cos 9$ to at least 10 decimal places. [Suggestion: Program your computer to handle multiple precision. You may start with .9 as a first approximation to $\cos 9$.]

4 Old Values of π

Let us consider some old values of π and how they may have been discovered.

3. The Biblical Value of π (*I Kings vii 23*) was common in many ancient cultures, as we saw in Chapter 5. As the ratio of the circumference to the diameter, there is nothing mysterious about this value. The roughest measurements of diameter and circumference with a tape measure will reveal the ratio to be a little greater than 3.

$\sqrt{10} = 3.16227766$. Again, it is not hard to imagine how this value might be obtained. Using a tape measure, the circumference of a circle a foot in diameter is just under 3 feet 2 inches, i.e. π is approximately $3\frac{1}{6} = \frac{19}{6} = 3.1\overline{6}$. If at some point one squares 19, one gets 361 and notices

$$\left(\frac{19}{6}\right)^2 = \frac{361}{36} \approx \frac{360}{36} = 10.$$

That $\sqrt{10}$ is actually a better approximation to π than 19/6 is either serendipitous or indicates that a better explanation of its use as a value of π can be found.

Here let me quote Petr Beckmann's *A History of Pi* [1]:

The Hindu mathematician Brahmagupta (born 598 A.D)[2] uses the value

$$\pi \approx \sqrt{10} = 3.16277\ldots$$

---

[1] p. 27. Bibliographical information is to be found in the Bibliography.
[2] The use of $\sqrt{10}$ as an estimate for π by Indian mathematicians is considerably older than this.

which is probably... based on Archimedean polygons. It has been suggested[3] that since the perimeters of polygons with 12, 24, 48 and 96 sides, inscribed in a circle with diameter 10, are given by the sequence

$$\sqrt{965}, \sqrt{981}, \sqrt{986} \text{ and } \sqrt{987},$$

the Hindus may have (incorrectly) assumed that on increasing the number of sides, the perimeter would ever more closely approach the value $\sqrt{1000}$, so that

$$\pi = \sqrt{1000}/10 = \sqrt{10}.$$

Start with a regular hexagon inscribed in a circle of radius 5 (diameter = 10) and at each stage double the number of sides of the inscribed figure.

i. Show that if a regular polygon inscribed in a circle of radius 5 has sides of length x, the next polygon will have sides of length

$$\sqrt{50 - 5\sqrt{100 - x^2}}.$$

ii. Show that, after n doublings, the perimeter of the inscribed polygon will be

$$P(n) = 6 \times 2^n \times s(n),$$

where

$$s(0) = 6 \cdot 5 = 30$$

$$s(n+1) = \sqrt{50 - 5\sqrt{100 - s(n)^2}}.$$

iii. Make a table of values of $P(n)^2$ for $n = 0, 1, \ldots, 10$ and verify the values quoted by Beckmann for $n = 1, 2, 3, 4$ (i.e., 12-, 24-, 48-, and 96-sided polygons). Why would it be natural to calculate $P(n)^2$ instead of $P(n)$?

iv. Today it is more natural to deal with circles of radius 1. Show that in a circle of radius 1, if one starts with an inscribed regular polygon of side x, and doubles the number of sides, the new polygon will have sides of length

$$\sqrt{2 - \sqrt{4 - x^2}}.$$

Define

$$s(0) = 1$$

$$s(n+1) = \sqrt{2 - \sqrt{4 - s(n)^2}}$$

---

[3] Here, Beckmann gives a footnote to p. 19 of Ferdinand Rudio, *Archimedes, Huygens, Lambert, Legendre. Vier Abhandlungen über Kreismessung*, Leipzig 1892. This explanation can, in fact, already be found in Hermann Hankel's *Zur Geschichte der Mathematik in Alterthum und Mittelalter*, B.G. Teubner, Leipzig, 1874, pp. 216 - 217.

and make a table of values of $P(n) = 6 \times 2^n \times s(n)$ for $n = 0, 1, \ldots, 10$.

22/7. The Archimedean estimate $\pi \approx 22/7$ was the most accurate value known to antiquity. Most improvements on this value before the advent of the Calculus were arrived at by iterating Archimedes' polygonal approximations to the circle beyond the 96-sided figures he used. Using the notation of our discussion above justifying the use of $\sqrt{10}$ as an approximation, we can say that he showed $P(4) < \pi$ and estimated further that

$$3 + \frac{10}{71} \approx 3.14084509 < 3.141031951 \approx P(4).$$

Looking at the circumscribed regular 96-gon, he obtained

$$\pi < 3.14280279 < 3.142857143 \approx \frac{22}{7}.$$

The perimeter of the circumscribed polygon is readily determined from that of the inscribed one. Look at *Figure 3*. If DF is the side of an inscribed n-gon, then AC will be the side of the circumscribed n-gon and one has, if one assumes a radius of 1,

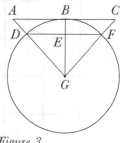

Figure 3

$$\frac{AC}{DF} = \frac{BC}{EF} = \frac{BG}{EG} = \frac{1}{EG}$$

$$= \frac{1}{\sqrt{1 - (EF)^2}} = \frac{2}{\sqrt{4 - DF^2}}.$$

For the 96-gon, the side DF is approximately .06544, whence

$$AC = .06544 \times \frac{2}{\sqrt{4 - (.06544)^2}} \approx .0654750581.$$

Multiplying by 96 yields a perimeter of 6.285605581, which is 2×3.14280279.

$4 \times \left(\frac{8}{9}\right)^2 \approx 3.160493827$. This is the Egyptian value of π. No one knows how the Egyptians arrived at this value, but I've seen several explanations, all relating π to the area of the circle rather than to the circumference. The simplest explanation is based on an illustration in the Rhind papyrus similar to *Figure 4* on the next page.

The interpretation is that the area of the circle of diameter 9 is being approximated by the area of the regular octagon of side 3. Now this area is the difference of the area of the square and the four triangular corners:

$$A = 9^2 - 4 \cdot \frac{1}{2} \cdot 3 \cdot 3 = 9^2 - 2 \cdot 9 = 63.$$

This is approximately 64, which is a perfect square. Thus, the area of a circle of diameter 9 is

$$\pi \left(\frac{9}{2}\right)^2 \approx 8^2$$

and we have

$$\pi \approx 4 \times \left(\frac{8}{9}\right)^2.$$

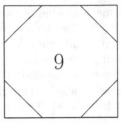

Figure 4

An explanation so simple one could present it to elementary school children again refers to a 9-by-9 square, but this time with the circle inscribed as in *Figure 5*.

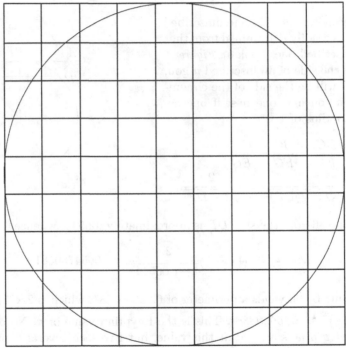

Figure 5

First, one counts the squares that are almost completely within the circle. There are 57 of these. Then one adds $4 \times \frac{3}{4}$ for the 4 squares that are approximately three-quarters covered by the circle. Finally, one adds $8 \times \frac{1}{2}$ for the 8 squares approximately half covered. This yields

$$57 + 4 \cdot \frac{3}{4} + 8 \cdot \frac{1}{2} = 57 + 3 + 4 = 64.$$

Thus $\pi \times \left(\frac{9}{2}\right)^2 \approx 8^2$.

In *The Crest of the Peacock*, George Gheverghese Joseph rejects the first of these explanations as "rather contrived and unconvincing" [4] and points to the popularity of geometric designs in ancient Egypt and cites P. Gerdes[5] for a configuration of 64 discs in a circle of radius approximately 9 discs in diameter. This could make a fairly convincing classroom demonstration using 64 small coins, perhaps sliding them around on the glass of an overhead projector: One starts with a single coin in the centre, tightly surrounding it by 6 adjacent coins. The next outer ring has 13 coins and is not a tight fit. Then comes a ring of 19 coins and finally one of 25.

There is yet another, albeit not too plausible approach. One starts with a given approximation to π such as 19/6, and makes a table

$$\frac{19}{6}\left(\frac{1}{2}\right)^2, \frac{19}{6}\left(\frac{2}{2}\right)^2, \ldots$$

of areas of circles of diameters $1, 2, \ldots$ until one finds a value close to a perfect square:

$$\frac{19}{6}\left(\frac{9}{2}\right)^2 = 64.125$$

and then replaces the near-square by the square and the approximation 19/6 to π by π to obtain a new approximation

$$\pi \times \left(\frac{9}{2}\right)^2 \approx 8^2.$$

v. Make a table of values of $n, \frac{19}{6}\left(\frac{n}{2}\right)^2$, and $\left[\sqrt{\frac{19}{6}\left(\frac{n}{2}\right)^2}\right]^2$ ($[\cdot]$ denoting the greatest integer function) and see for which values of n the entries in the latter two rows are close. Use these to approximate π.

vi. Do the same with 19/6 replaced by $\sqrt{10}$.

vii. Do the same with 19/6 replaced by the Egyptian value for π. What do you notice?

viii. Let p be an initial approximation to π and let

$$p' = \frac{4}{n^2}\left[\sqrt{p}\frac{n}{2}\right]^2.$$

Show that $p' \leq p$ and thus the procedure we've been using cannot improve our estimate for π if $p < \pi$.

---

[4] pp. 83 - 84.

[5] "Three alternative methods of obtaining the ancient Egyptian formula for the area of a circle", *Historia Mathematica* 12 (1965), pp. 261 - 268.

5 Using Polynomials to Approximate π

Once we know Calculus, π is readily determined by infinite series. One can also use some simple approximation techniques. Passage from an inscribed n-gon to an inscribed $2n$-gon can be viewed as replacing the circular excesses by triangles. One could as well use parabolas.

Consider *Figure 6*, in which a square of area $\sqrt{2} \times \sqrt{2} = 2$ is inscribed inside a circle of radius 1.

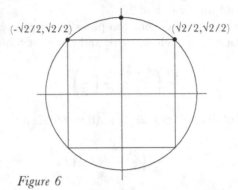

Figure 6

i. Estimate the area of the region between the circle and the top of the square by fitting a parabola $Y = aX^2 + bX + c$ to the points $(-\sqrt{2}/2, \sqrt{2}/2)$, $(0,1)$, and $(\sqrt{2}/2, \sqrt{2}/2)$ and evaluating the integral

$$A = \int_{-\sqrt{2}/2}^{\sqrt{2}/2} \left(y - \frac{\sqrt{2}}{2} \right) dx.$$

The area of the circle, and hence π, is approximately $2 + 4A$.

ii. Use the points $(\sqrt{2}/2, \sqrt{2}/2)$, $(\sqrt{3}/2, 1/2)$, and $(1,0)$ to find another parabola to use to estimate the area A_2 of the upper half of the region trapped between the circle and the right side of the square. This yields $\pi \approx 2 + 8A_2$, or even $\pi = 2 + 2A + 4A_2$.

iii. Use the points $(0,1)$, $(\sqrt{2}/2, \sqrt{2}/2)$, and $(1,0)$ to give a new estimate for A_2.

Of course, one need not restrict onself to parabolas. A good pocket calculator, like the *TI-83+*, will perform a variety of statistical regressions. Five points on the upper semi-circle will uniquely determine a 4th degree polynomial, the coefficients of which can be determined by Quartic Regression.

iv. Choose 5 points on the unit circle in the first quadrant and apply a quartic regression to obtain a polynomial approximation. Graph the polynomial and integrate it between 0 and 1. This should approximate $\pi/4$. Repeat this with 5 different points.[6]

---

[6] You might try $0°, 30°, 45°, 60°$, and $90°$ first, and then try $0°, 45°, 60°, 75°, 90°$ or $0°, 15°, 30°, 45°, 90°$ to see the effect of clustering one's points near the extremes.

v. Use the entire semicircle, choosing $0°, 45°, 90°, 135°, 180°$. Doubling the area in this case yields $\pi \approx 3.108494467$ on the *TI-83+*. Add $30°, 60°, 120°, 150°$ to the list of angles before performing the quartic regression. This should fit the circle better. What value does this yield for π? Plotting a point every 15 degrees from 0 to 180 yields $\pi \approx 3.127222704$,[7] not as good as $22/7$, but a respectable value nonetheless.

For those who possess a *TI-83+*, but are not yet familiar with its regression functions, let me outline briefly how to do the above. Start in the List Editor with empty lists L_1, L_2, L_3. Enter several values for angles in L_1 Alternatively, one can do this before entering the Editor. In the last part of task *v*, for example, I generated L_1 in the main screen by entering

$$\mathrm{seq}(X, X, 0, 180, 15) \rightarrow L_1.$$

The List function seq takes an expression and variable as its first two inputs, a starting and a stopping point as its next two, and an optional increment as a fifth input. Thus, if I wanted to, I could have generated the same list by typing in

$$\mathrm{seq}(15X, X, 0, 12) \rightarrow L_1,$$

the increment in X now being 1 by default.

Back in the Editor, one positions the cursor above the list entries in the L_2 column and enters the expression

$$"\cos(L_1)".$$

Similarly, one enters

$$"\sin(L_1)"$$

for the L_3 column. As soon as one hits ENTER, the entries are generated. [One can also do this in the main screen by entering

$$"\cos(L_1)" \rightarrow L_2$$

$$"\sin(L_1)" \rightarrow L_3,$$

successively. Of course, if only one list L_1 of angles is to be used, one can dispense with the quotation marks and simply enter

$$\cos(L_1) \rightarrow L_2 \quad \text{and} \quad \sin(L_2) \rightarrow L_3.$$

Now that one has generated the lists, it is time to do the regression. One goes in the CALC submenu of the Stat menu and chooses QuartReg. Back on the main screen one now sees

---

[7] On the *TI-83+*, integration in graphing mode is set to a tolerance of .001, while for integration using the *fnInt* item in the Math button menu the tolerance is set to .00001. I have here used the former option.

QuartReg

and completes the expression by choosing L_2, L_3 and a variable to store the equation in. For example,

$$\text{QuartReg } L_2, L_3, Y_1.$$

The variable Y_1 is found by pressing the VARS button and navigating to the Y-VARS submenu and thence to the function submenus of that. Y_1 is the first choice on the list.

When you hit ENTER, the coefficients of the quartic approximation to the unit semicircle appear on the screen. If you want, you can now integrate them by hand between the values -1 and 1. Or, given that the expression is now stored in Y_1 in the list of functions, you can simply graph the function and let the calculator determine the area in the manner I assume already familiar to you. Bear in mind, however, that integration stops when the built-in tolerance is achieved[8].

Of course, with so many points graphed, there is no reason to stick with a quartic regression other than that the *TI-83+* doesn't have any buttons for polynomial regressions of higher order. However, the *TI-83+* and other calculators dealing with lists make the calculations of the coefficients of higher degree regression polynomials a snap and one might like to see how the value of π changes with a given set of plotted points as one varies the degrees of the regression polynomials.

6 π à *la* Horner

Apply Horner's Method (specifically, formula (17) of Chapter 7) to

$$f(x) = \sin\left(\frac{x}{6}\right) - \frac{1}{2}$$

to approximate π. Do the same with the Newton-Raphson method and compare the workings of the two procedures.

---

[8] Cf. footnote 7. On the *TI-85*, one can reset the tolerance to achieve greater accuracy. Some newer calculators, like the *TI-89*, do symbolic integration and presumably the tolerance issue can be avoided on them, at least for simple polynomials. Lest it begin to seem like this book is becoming an advertisement for Texas Instruments^{TM}, let me hasten to add that I refer to their calculators because in the two most recent schools I've worked at the *TI-85* and *TI-83+* were the respective calculators the Math Departments had chosen to standardise instruction on, and thus are the calculators I am familiar with.

7 Parabolas

The history of circle measurement is quite involved, with multiple origins in ancient cultures lost to us, and several distinct approximations to π, each with its own set of plausible scenarios of discovery. The origins of the parabola are not as shrouded in mystery. Report on the techniques of finding the areas under a parabolic segment through the ages.

8 Finite Geometries and Bradwardine's Conclusion 38

From a modern point of view, geometry is a branch of algebra and not the theory of space. One of the less abstract approaches is to regard n-dimensional geometrical space as an n-dimensional vector space over the reals. Replacing the reals by other fields gives us alternative geometries which obey some, but not all the Euclidean axioms.

In the following, p will denote an arbitrary fixed prime number. Let

$$\mathbb{Z}_p = \{0, 1, \ldots, p-1\}$$

and define operations $\oplus$ and $\odot$ on $\mathbb{Z}$ by

$$x \oplus y = \text{ remainder of } x + y \text{ after dividing by } p$$
$$x \odot y = \text{ remainder of } x \cdot y \text{ after dividing by } p,$$

where $+, \cdot$ are the usual operations on the integers. In a Modern Algebra or a Number Theory course, one proves that $\{\mathbb{Z}_p; \oplus, \odot, 0, 1\}$ is a field: for all $x, y, z \in \mathbb{Z}_p$,

$$x \oplus y = y \oplus x \qquad\qquad x \odot y = y \odot x$$
$$x \oplus (y \oplus z) = (x \oplus y) \oplus z \qquad x \odot (y \odot z) = (x \odot y) \odot z$$
$$x \oplus 0 = x \qquad\qquad x \odot 1 = x$$
$$\exists w (x \oplus w = 0) \qquad\qquad x \neq 0 \to \exists w (x \odot w = 1)$$
$$x \odot (y \oplus z) = (x \odot y) \oplus (x \odot z).$$

Once one has shown this, one no longer needs to distinguish $\oplus$ from $+$ and $\odot$ from $\cdot$ and the simpler notation is invoked.

The *points* of the *p-plane* are just the ordered pairs of elements of $\mathbb{Z}_p$:

$$\mathbb{P}_p = \{(x, y) | x \in \mathbb{Z}_p \text{ and } y \in \mathbb{Z}_p\}.$$

The *lines* of the p-plane are the solutions sets in $\mathbb{P}_p$ to linear equations,

$$aX + bY + c = 0, \qquad a, b, c \in \mathbb{Z}_p.$$

If $(x, y) \in \mathbb{P}_p$ satisfies $ax + by + c = 0$, then (x, y) is said to lie on the line defined by $aX + bY + c = 0$.

i. Show: For any two distinct points of a p-plane there is an unique line in the plane on which the two points lie.

ii. Show: The equation of any line can be uniquely written in one of the forms,

$$X = c$$
$$Y = mX + b, \tag{2}$$

with $m, b, c \in \mathbb{Z}_p$.

If the equation of a line can be written in the form (2), the number m is called the *slope* of the line. As usual, a line is *horizontal* if its equation can be written in the form $Y = b$, and *vertical* if it can be written in the form $X = c$. Two lines are *parallel* if they are both vertical or if they have the same slope.

iii. Show: a. Parallel lines do not intercept.

b. Given any line L and a point P not on the line, there is an unique line L_P parallel to L and on which P lies.

c. Two distinct non-parallel lines intersect in an unique point.

Perpendicularity can also be defined, but is is a bit more problematic. We say two intersecting lines L_1, L_2 are *perpendicular* if either one of L_1, L_2 is horizontal and one vertical, or the lines have slopes m_1, m_2 where $m_1 \cdot m_2 = -1$ (i.e. $m_1 \cdot m_2 = p - 1$).

iv. Show: For $p = 5$, the line $Y = 2X$ is perpendicular to itself.

v. Show: Given any line L not perpendicular to itself and a point P not on the line, there is an unique line L_P perpendicular to L on which P lies.

A few basic facts about the p-plane are these:

vi. Show: Every line has p points.

vii. Show: Every point lies on $p + 1 = \frac{p^2 - 1}{p - 1}$ lines.

viii. Show: The plane has p^2 points and $p^2 + p$ lines.

Because of the cyclic nature of $\mathbb{Z}_p$, there is not an ordering of the points on a line and it is not clear what it means for one point to be adjacent to another. However,

ix. Show: If every point is adjacent to only two points on each line, then the number of points in the plane adjacent to a given point is $2(p + 1)$.

Harclay's assumption that all continua consist of adjacent points compels us to assume that every point on a line is adjacent to one or more such points— our familiarity with the Euclidean plane suggests in fact that each point on the line is adjacent to exactly 2 such points. How do we find the adjacent points and the number of them in the p-planes? Which p-planes have exactly 2 points adjacent to a point on every line the point lies on?

A point (x, y) in the real plane is called a *lattice point* if x, y are both integral. We can graphically represent a line in the p-plane by identifying the points of the p-plane with the lattice points of the real plane and identifying the points on the lines in the p-plane with the lattice points of the corresponding line in the real plane— all the while reducing modulo p. For example, *Figure 7* represents the line $Y = 3X$ for $p = 7$, bearing in mind that

$$y = \begin{cases} 3x, & 0 \le 3x < 7 \\ 3x - 7, & 7 \le 3x < 14 \\ 3x - 14, & 14 \le 3x < 21. \end{cases}$$

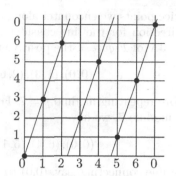

Figure 7

It seems clear that (0,0) is adjacent to (1,3) and, in the opposite direction, to (6,4).

x. Copy *Figure 7* to some graph paper. On the same paper graph the lines $Y = \frac{2}{3}X, Y = 7 - \frac{1}{2}X$ and $Y = 7 - \frac{1}{4}X$. What points are adjacent to (0,0) on these lines? How many points are adjacent to (0,0) on $Y = 3X$?

xi. Let $p = 2n + 1$ be an odd prime greater than 3. Show: On the line $Y = nX$ there are more than 2 points adjacent to (0,0). [Hint: Consider $Y = \frac{1}{2}X$.]

xii. Let p be a prime number greater than 3. Show: The number of points adjacent to a given point is greater than $2(p + 1)$.

xiii. Let $p = 2$. Show: Every point is adjacent to 3 other points.

xiv. Show: if $p = 3$, then there are exactly two points adjacent to a given point on any line the point lies on. Deduce Conclusion 38: In the 3-plane every point has exactly 8 immediate neighbours.

We can do better than this. Notice that every "non-vertical" line in the p-plane passing through (0,0) has a representation $Y = mX$ for some integer $-n \le m \le n$, where $p = 2n + 1$. The real lines

$$Y = mX + kp, \qquad k = 1, 2, 3, \ldots$$

define parallel lines in the real plane, but all define the same line in the p-plane. Similarly, the vertical lines $X = kp$ for $k = 1, 2, 3, \ldots$ are parallel to $X = 0$ and all define the same line in the p-plane as $X = 0$.

xv. Let P, Q be points on the lines $Y = mX$ and $Y = mX + p$, respectively, with P, Q not identified in the p-plane. Show: The line L_p in the p-plane consisting of the lattice points of the line L connecting P, Q coincides with the line defined by $Y = mX$. Moreover, there are no lattice points on L between P and Q.

xvi. Repeat xv for the vertical lines $X = 0$ and $X = p$. Conclude: (0,0) is adjacent to every point other than (0,0) on the lines $Y = mX$ and $X = 0$ in the p-plane.

It follows that every point of the p-plane is adjacent to every other point.

Before moving on, let me offer some motivation. Take any line L in the real plane passing through the origin. In the real plane, the line has only two directions proceeding away from (0,0). If we follow one of these, and list the lattice points we encounter along the way, we can consider this list as defining a direction for the line considered in the p-plane. For example, for the line $Y = 3X$ of *Figure 7*, the positive real direction yields

$$(0,0), (1,3), (2,6), (3,2), (4,5), (5,1), (6,4)$$

before repeating modulo $p = 7$. Following the negative direction, after reducing modulo 7, we have

$$(0,0), (6,4), (5,1), (4,5), (3,2), (2,6), (1,3).$$

The line connecting, say, (0,0) to (4,5) coincides in the p-plane with $Y = 3X$ and results in

$$(0,0), (4,5), (1,3), (5,1), (2,6), (6,4), (3,2).$$

It turns out that each line in the p-plane has $p - 1$ directions and hence each point on the line has $p - 1$ adjacent points on the line.

xvi. Define a circle of radius k around a point P in the p-plane to be the set of all points reachable from P by proceeding k steps in all directions along all lines passing through P. Show: If p does not divide k, the circle of radius k consists of all points other than P, while, if p divides k, the circle consists of P alone.

9 Root Extraction

Look up the old Chinese or Hindu methods of root extraction for square roots and cube roots. How did Theon of Alexandria extract square roots? Finally, look up Horner's original paper and explain his method of finding cube roots. Apply the Chinese or Hindu method and the Newton-Raphson method to the example given by Horner and compare the workings of these procedures with Horner's.

10 Statistical Analysis

Compile a list of names of mathematicians from the index of any standard History of Mathematics textbook. Then go to a good research library and select a number of general reference works including the *Dictionary of Scientific Biography*, the *Encyclopedia Britannica*, the *World Who's Who of Science*, the *Encyclopedia Americana* a few foreign language encyclopædias, and perhaps the *New Catholic Encyclopedia* and the *Encyclopædia Judaica*. Construct a table of birth and death dates given for the mathematicians chosen. For each individual, pick a "correct" pair of dates by choosing the most popular one. If no clear winner emerges, declare an experimental error and delete the individual from the list. For each reference work, calculate *i.* how many names from your list have entries in the work;

    ii. how many entries have the correct birth dates;

    iii. how many entries have the correct death dates.

Compare the reference works with respect to breadth of coverage and reliability.

11 The Growth of Science

Choose a volume of the *Dictionary of Scientific Biography*. Construct a table as follows. In the first column list the centuries from -5 (500 - 401 B.C) to 20 (1901 - 1972). In the second column enter the number of individuals who died or, if there is no death date, flourished in that century. In the third column enter the number of such scientists receiving more than 2 pages of coverage, and in the fourth column the number of such receiving 5 or more pages. Analyse your results.

    An alternative to this is to go to the library and jot down year-by-year the total number of pages in *Mathematical Reviews*, *Zentralblatt für Mathematik*, or the Russian review journal.

    Yet another possibility is to choose a chronology such as Darmstædter's or Parkinson's books cited in the Bibliography and count the number of entries per year, decade, or century given in the book and analyse these results.

    In what ways are these good samplings of mathematical activity? In what ways might they not be?

12 Programming

In writing Chapter 7, I saved myself a lot of computation by writing programs for my calculator to multiply polynomials (entered as lists of coefficients) and to determine the coefficients of the polynomial obtained by performing the substitution $X = Y + a$. Write a suite of programs for your calculator (or, say, in LOGO) powerful enough to handle the problems discussed in Chapter 7. Include provisions for multiple precision and match Mr. W. Harris Johnston of Dundalk (page 184) by solving Newton's cubic equation to 101 decimal places.

Index

ϕ, 52–59
π, 36, 67, 71, 89, 136, 137, 142, 233, 249–257

Aaboe, Asger, 24
Abel, Niels Henrik, 130–131, 242
Abir-Am, P.G., 30, 31
Ader, Clément, 12
Airy, George Biddel, 232
Ajram, K., 17
Albert, Abraham Adrian, 200
Alembert, Jean Lerond d', 240
Alexander of Aphrodisias, 50
Alexandrov, Pavel Sergeivich, 35
Alfonso X, 230
Algazel, see Ghazzālī, al-
algebraic
    curve, 100, 102
    integer, 163, 165
    number, 118
Alic, Margaret, 31
Archimedean Axiom, 60, 61, 64, 68
Archimedes, 2, 38, 60, 68, 229, 238, 251
Aristarchus, 17, 25
Aristotle, 5, 50, 52, 69–70, 72, 73, 243
Arnold, H., 244
Arnold, Lois Barber, 32
Artin-Schreier Theory, 224
Asimov, Isaac, 21
Auer von Welsbach, Carl, 11–12, 241
Averroës, see Rushd, ibn
Avicenna, see Sina, ibn

Babbage, Charles, 33, 232, 243

Bachet de Meziriac, Claude Gaspar, 13
Bailey, Martha J., 31
Banach, Stefan, 234
Banneker, Benjamin, 242
Banting, Frederick, 233
Baron, Margaret E., 29, 176, 224
Bartolozzi, M., 197, 199, 215
Barton, Otis, 234
Baum, Joan, 33
Beckmann, Petr, 36, 249
Bedient, Jack D., 24
Beebe, Charles William, 234, 235
Bell, Eric Temple, 6, 14, 20, 22, 233
Berggren, John Lennart, 27, 36
Bernoulli, Jakob, 143, 231
Bernoulli, Johann, 137, 143, 231
Best, Charles Herbert, 233
Bhaskara, 43
Bieberbach, Ludwig, 89
Binomial Theorem, 178, 181, 227
Bishop, W.J., 38
Blake, William, 26
Bolyai, János, 29, 231, 244–246
Bolzano, Bernhard, 16–17
Bombelli, Rafaello, 13, 17–18
Bonola, Robert, 29
Borwein, Jonathan, 36
Borwein, Peter, 36
Bougainville, Louis Antoine de, 243
Bourbaki, Nicolas, 7
Boyer, Carl, 22, 29
Bradwardine, Thomas, 8, 9, 72–76, 82–85, 257

Brahmagupta, 249
Brewster, David, 28
Brouwer, Luitzen Egbertus Jan, 234
Bucciarelli, Louis I., 34
Buchner, P., 96
Budan, Ferdinand François Désiré, 183–184, 199
Buddha, 2
Bunt, Lucas N.H., 24
Burton, David, 3, 66, 133, 135–136, 176–177
Byron, Ada, see Lovelace, Ada

Cajori, Florian, 11, 21, 69
Calinger, Ronald, 26
Camp, Bob, 246
Campbell, Douglas M., 26
Campbell, George, 215–217, 224
Cantor, Georg, 37, 232
Cantor, Moritz, 11, 14, 21, 233, 236
Cardano, Girolamo, 198
Carlisle, Thomas, 28
Carroll, Lewis, 3, 28
Cauchy, Augustin Louis, 16–17, 30, 182
Ch'in Chiu-shang, see Qín Jiǔsháo
Ch'in Chiu-shao, see Qín Jiǔsháo
Ch'in Chu-shao, see Qín Jiǔsháo
Chabert, Jean-Luc, 26, 181–182
change of signs, 197
Chebyshev, Pafnuty Lvovich, 13
Cheng Heng, 137
Chhin Chiu-Shao, see Qín Jiǔsháo
Chin Kiu-shao, see Qín Jiǔsháo
Chu Shih-chieh, 176
Clagett, Marshall, 73
Clairaut, Alexis Claude, 240
Cohen, Morris R., 5, 25
Condorcet, Antoine Nicolas Marquis de, 231
Confucius, 2
constructible
    circle, 92
    number, 91
    point, 92
    straight line, 92
constructibly determined polynomial, 102
Cooke, Roger, 34

Coolidge, Julian Lowell, 10, 20, 29, 184–187, 189–195, 223–224
Cooney, Miriam, 32
Copernicus, Nicolaus, 17, 225, 239
Courant, Richard, 80, 89
cubic resolvent, 98–99, 109, 111
Curie, Marie, 32, 225
Curtze, Maximilian, 73

Däniken, Erich von, 35
Darmstædter, Ludwig, 37, 261
Darwin, Charles, 20, 231
Dauben, Joseph, 15, 41, 43
David, F.N., 35
David, L. von, 244
Davies, T.S., 185
Davis, Martin, 35
de Gua de Malves, Jean Paul, see Gua de Malves, Jean Paul de
de Gua point, 217–222
    multiplicity of, 221
de Gua triple, 217
    order of, 221
De Gua's Theorem, 8, 10, 197, 211–212, 214–222, 224
Debus, Allen G., 19
Dedekind, Richard, 4, 68
Dee, John, 231
Dekker, Douwes, 43
Democritus, 69, 73
DeMoivre's Theorem, 168
DeMoivre, Abraham, 168
Descartes' Rule of Signs, 8, 10, 187, 189, 196–215, 224
Descartes, Rene, 76, 80, 87, 105, 196, 198–199, 240
Dick, Auguste, 35
Dickson, Leonard E., 131, 200
dihedral angle, 82, 247
Dingler, Hugo, 238
Diophantus of Alexandria, 12–14, 17–18, 37
discriminant, 127, 148, 151, 156, 168, 171
divine proportion, see golden ratio
Drabkin, I.E., 5, 25
Dù Shíràn, 27, 41, 137–139, 145, 177, 179, 199
Dürer, Albrecht, 20

Duns Scotus, John, 72, 76
duplication of the cube, 87, 92–93
Dworsky, Nancy, 34
Dzielska, Maria, 32

Eddington, Arthur Stanley, 236
Edwards, C.H., Jr., 30
Einstein, Albert, 233, 242
Eisenstein's Criterion, 116–118
Euclid, 1–7, 15, 24–26, 28, 37, 38, 42–48,
    50–51, 60, 62–68, 72–74, 77, 79,
    88, 92, 96, 111, 115, 231
Euclid of Megara, 2, 38
Euclidean geometry, 92
Eudoxus, 2, 4, 6, 52, 59–60
Euler's formula, 80–82
Euler, Leonhard, 5, 80, 186–187, 189,
    212, 214, 240
Eurytus, 50
Eves, Howard, 22, 25

Farnes, Patricia, 31
Fáry, István, 245
Fauvel, John, 26
Fermat, Pierre de, 12, 235
Fermat, Samuel de, 12, 17
Feynman, Richard, 242
Fibonacci, 133, see Leonardo of Pisa
field
    finitely generated number, 124
    number, 89
finite differences, 190
Flammarion, Camille, 11, 241
Fourier, Jean Baptiste Joseph, 184, 199,
    202
Franci, R., 197, 199, 215
Freud, Sigmund, 232
Freudenthal, Hans, 16
Friedlein, G., 5
Fundamental Theorem of Algebra, 208

Gaal, Lisl, 35, 131
Gagarin, Yuri, 235
Galen, 231
Galileo, 18, 240
Galois group, 129
Galois Theory, 87–89, 99, 129–131
Galois, Evariste, 130–131

Gauss, Carl Friedrich, 88, 111, 198–200,
    202, 210–211, 216, 231, 236, 244,
    246
Gellert, W., 244
Gentzen, Gerhard, 234
geometric algebra, 16, 26
Gerbert of Aurillac, 243
Gerdes, P., 253
Germain, Sophie, 34
Ghazzālī, al-, 70–73, 76
Gibbs, Josiah Willard, 242
Gibson, Bob, 246
Gilbert, Davies, 175, 184
Gillespie, Charles, 20
Gillings, Richard, 24
Gittleman, Arthur, 44, 50
Gödel, Kurt, 15–16, 26
golden ratio, 36, 48, 52, 55, 57, 59
Grabiner, Judith V., 30
Grant, Edward, 25, 70, 72, 73
Grattan-Guinness, Ivor, 15–16, 38, 245
Gray, Jeremy, 20, 26, 223, 246
Green, Judy, 31
Greenberg, Marvin Jay, 29, 245
Greenwood, Thomas, 50
Grosseteste, Robert, 73
Gua de Malves, Jean Paul de, 187, 189,
    199, 202, 212, 214, 224

Hadlock, Charles Robert, 89, 125
Hankel, Hermann, 85, 250
Harriot, Thomas, 176, 198–199
Hartshorne, Robin, 87
Hawking, Stephen, 26
Heath, Thomas, 24, 26, 44
Heiberg, J.L., 38
Hellwich, M., 244
Henle, James, 1–4, 6, 15
Henricus Modernus, see Henry of
    Harclay
Henry of Harclay, 72, 73, 75, 76, 85, 258
Henry, John, 246
Heron, 68
Herstein, Israel Nathan, 131
Hesse, Otto, 102
Higgins, John C., 26
Hilbert Irreducibility Theorem, 122,
    124, 126, 128
Hilbert's 17th Problem, 224

Hilbert, David, 4, 13, 65–66, 234, 238
Hippassus, 52
Hippias, 87
Ho Peng-Yoke, 135, 136, 138, 145, 146
Homer, 243
Horner's Method, 7–9, 137, 175–197,
    222–224
Horner, William George, 10, 175–196,
    212, 214, 216–217, 221, 223, 230,
    241, 249, 260
Hypatia, 31, 32
hyperbolic functions, 170

Imhotep, 229
incommensurability, 50
Intermediate Value Theorem, 158, 200,
    202, 207–208, 216
irreducible
    curve, 101
    polynomial, see polynomial,
        irreducible

Jacobson, Nathan, 224
Johnston, W. Harris, 184, 261
Jones, Phillip S., 24
Jordan, Camille, 130
Joseph, George Gheverghese, 27, 41,
    177, 179, 223, 253

Karman, Theodore von, 242
Karpinski, Louis Charles, 28
Kass-Simon, G., 31
Kästner, H., 244
Katz, Victor J., 60, 63, 156, 176–177,
    245
Kendall, Phebe Mitchell, 227
Kepler's Laws of Planetary Motion,
    226, 240
Kepler, Johannes, 5, 225–226, 239
Khayyami, al-, 87, 105
Khwarezmi, al-, 12, 37, 133
Klein, Felix, 29, 87–89, 122, 130–131,
    232
Klein, Jacob, 35
Kline, Morris, 23
Kneller, Godfrey, 226–229
Knorr, Wilbur Richard, 29
Koblitz, Ann Hibner, 34
Kochina, Pelageya, 34

Kornbluth, Cyril, 236
Korolev, Sergei P., 226
Kovalevskaya, Sofya, 34–35
Krishnaiah, P.V., 200, 203, 224
Kronecker, Leopold, 13, 232, 234
Küstner, H., 244

Ladies' Diary, 175, 185–189, 223
LaDuke, Jeanne, 31
Lagrange, Joseph Louis, 189, 200,
    212–213, 240
Laguerre, Edmond Nicolas, 199
Lakatos, Imre, 80
Landau, Edmund, 89
Laplace, Pierre Simon de, 240
Lasswitz, Kurt, 236, 244
lattice point, 258
Laugwitz, Detlef, 89, 96–97, 106, 131
Laurus, T.B., 239
Law of Cosines, 46
Law of Gravitation, 226
Least Number Principle, 55
Lebesgue, Henri, 26
Lefkowitz, Mary, 17
Legendre, Adrien Marie, 28
Leibniz, Gottfried Wilhelm, 199
Leonardo da Vinci, 20
Leonardo of Pisa, 230
Levin, Stewart A., 199
Li Chih, see Lĭ Yĕ
Li Juan, see Lĭ Rui
Lĭ Rui, 199
Lĭ Yan, 27, 41, 137–139, 145, 177, 179,
    199
Lĭ Yĕ, 137–138, 143, 145, 146, 160, 162
Li Yeh, see Lĭ Yĕ
Lĭ Zhì, see Lĭ Yĕ
Libbrecht, Ulrich, 142, 143, 145
Lindemann, Ferdinand, 36, 87–89, 233
Lister, Joseph Jackson, 242
Lobachevsky, Nikolai Ivanovich, 29, 246
Loomis, Elisha S., 36, 42
Lovelace, Ada, 33–34
Lützen, Jesper, 129

Macleod, J.J.R., 233
Maier, Anneliese, 73
Malves, Jean Paul de Gua de, see Gua
    de Malves, Jean Paul de

Marić, Mileva, 242
Matheson, N.M., 38
Matijasevich, Yuri, 35
maxima/minima, 98, 106, 154, 156, 216, 220, 222
Maziarz, Edward, 50
McClintock, Barbara, 242
McGrayne, Sharon Birch, 31
Mean Value Theorem, 194
Menæchmus, 3, 87, 105
Mendeleev, Dmitri, 225
Menninger, Karl, 28
Menzler-Trott, Eckart, 36, 83
Meschkowski, Herbert, 13
Method of Infinite Descent, 54
Mikami, Yoshio, 27, 41, 136–137, 145, 199
Millay, Edna Saint Vincent, 42
Miller, George Abram, 11
Mitchell, Maria, 32, 227
Mittag-Leffler, Gösta, 242
Mittag-Leffler, Charlotte, 34
Moore, Doris Langley, 33
Moore, Gregory H., 4
Morgan, Augustus de, 38, 175–176, 184, 238
Morgan, Sophia de, 38, 175
Morrison, Tony, 35
Morrow, Charlene, 32
Mouraille, Jean-Reymond, 182–183
Mozans, H.J., 30
Multatuli, see Dekker, Douwes
Murdoch, John E., 73–74, 83

Naber, H.A., 41–48
Napoleon, 12, 243
Nasr, Seyyed Hossein, 27
Nddeham, Joseph, 145
Needham, Joseph, 27, 135, 143
Netto, Eugen, 236–237, 241
Neugebauer, Otto, 17, 23
Neumann, John von, 242
Newman, James R., 3, 25–26
Newton's Method, 180–183, 193
Newton, Isaac, 26, 178, 180–184, 191, 193–194, 199, 212, 214, 225–229, 233, 240, 241, 261
Newton-Raphson Method, 181
Nicomedes, 87

Nightingale, Florence, 32, 232
Noether, Emmy, 32, 35

Ogilvie, Marilyn Bailey, 31
Osen, Lynn, 32
Oughtred, William, 176
Outram, D., 30

Pai Shang-shu, 143
Palter, Robert, 17
Pan Shang-shu, 144
Pappus, 25
Paracelsus, 231
Parkinson, Claire L., 37, 261
Pasbergius, Manderupius, 239
Pascal, Blaise, 20, 227
Pasch, Moritz, 65
pentagram, 55
Pepys, Samuel, 225
Perl, Teri, 32, 189
permanence of signs, 197
Petersen, Julius, 87–89, 100–103, 105–106, 122, 125, 128–130, 223–224
Piccard, August, 235
Pierpont, James, 87–89, 122, 130
Planck, Max, 12
Plato, 2, 5–6, 20, 48, 52, 59, 73, 237
Poggendorff, Johann Christian, 19
Pólya, Georg, 49, 199
polynomial
    elementary symmetric, 172
    irreducible, 102, 122–128
    minimal, 119
    prime, 119
    reduced, 98–99
    symmetric, 172
Pope, Alexander, 26
Prestet, Jean, 199
Proclus, 3–5, 7, 24, 65
Ptolemy, Claudius, 2, 17, 239
Pycior, H.M., 31
Pythagoras, 2, 7, 9, 36, 41, 48, 49, 73, 229
Pythagorean Theorem, 14–15, 36, 41–52, 134, 138–141

Qín Jiǔsháo, 133–146, 160, 175–176, 179–180, 182, 188, 190, 223, 230 quintisection, 111

Rademacher, Hans, 80
Raphael, 2
Raphson, Joseph, 181, 183
Raubdruck, 19
reduced
  cubic, 105
regular pentagon, 111
regular polygon, 77, 88, 95–96
Reichardt, Hans, 244–245
Reiche, Maria, 35
Reid, Constance, 35
Remane, Horst, 39
Rembrandt, 243
Rivlin, Lydia, 229–230
Robbins, Herbert, 80, 89
Robinson, Julia, 30, 35
Robinson, Raphael, 35
Rolle's Theorem, 158–160, 201, 222
rope stretchers, 14
Rudio, Ferdinand, 250
Ruffini, Paolo, 130–131, 175, 183–184
Rushd, ibn, 83
Russell, Bertrand, 4

Saccheri, Girolamo, 74
Sayers, Dorothy, 13
Schaaf, William L., 39
Schopenhauer, Arthur, 42
Schreiber, Peter, 39, 245
Schubert, Hermann, 236, 238
Segner, Johann Andreas, 199, 202, 215
Shanker, S.G., 130–131
Shepherdson, John, 54
Sina, ibn, 70, 229, 230
Skolem, Thoralf, 242
Smith, David Eugene, 22, 25, 27–28, 38,
    175, 177, 186, 188, 223
Socrates, 2, 17, 243
Solon, 2
Spinoza, Baruch, 65
squaring the circle, 87
Stack, N.G., 31
Stamm, Edward, 73, 75
Stein, Dorothy, 33
Struik, Dirk, 5–6, 22, 25
Sturm's Theorem, 224
Sturm, Charles François, 199

Sylow, Ludvig, 129, 131
Sylvester, James Joseph, 199
Sylvestre II, see Gerbert of Aurillac
synthetic division, 177–179
Szegö, Gabor, 199

Tarski's Decision Procedure, 224
Tarski, Alfred, 224, 234
Tartaglia, Nicolo, 162
Tartaglia-Cardano solution, 163
Taylor's Theorem, 176, 179, 182, 200,
    216, 218
Taylor, E.G.R., 21
Tchebichev, see Chebyshev
Thales, 2, 25
Theætetus, 5–6
Theodorus, 52
Theon of Alexandria, 260
Thomas, Ivor, 25
Thue, Axel, 242
TI-83+, 161, 164–165, 167, 195, 254–256
TI-85, 195, 256
TI-89, 256
Toeplitz, Otto, 80
Toole, Betty Alexandra, 33–34
Trigonometric Addition Formulæ, 94,
    111
trisection of the angle, 87, 94, 111
Truman, R.W., 39
Tschebyschew, see Chebyshev
Turing, Alan, 26
Tycho, 239

Unguru, Sabetai, 16

Vandermonde determinant, 171–172
Viète, François, 13, 17–18, 37, 168–169,
    176
Vogelweide, Walther von der, 243

Wærden, Bartel van der, 14, 16, 23, 35,
    99, 131
Wallis, John, 176
Waltherus Modernus, 73
Wāng Lái, 199
Wantzel, Pierre Laurent, 87–89, 122,
    130–131
Weber, Robert L., 39
Weierstrass, Karl, 36
Weil, André, 13, 16

Westfall, Richard S., 227–228
Weyl, Harmann, 35
Whiteside, D.T., 181
Wiener, Norbert, 246
Wiles, Andrew, 235
Wilson, Robin J., 40, 245
Witt, Jan de, 243
Wittgenstein, Ludwig, 243
Wussing, Hans, 39, 41, 244–245

Xylander, Guilielmus, 13

Yost, Edna, 31
Young, Edward, 228
Young, J.R., 176
Youschkevitch, A.P., 83, 145

Zeno, 69, 70
Zermelo, Ernst, 4, 15–16
Zoubov, V.P., 73